Transmission Line Reliability and Security

Transmission Line Reliability and Security

Anthony J. Pansini, EE, PE

River Publishers

Routledge
Taylor & Francis Group

LONDON AND NEW YORK

Published 2020 by River Publishers

River Publishers
Alsbjergvej 10, 9260 Gistrup, Denmark
www.riverpublishers.com

Distributed exclusively by Routledge

4 Park Square, Milton Park, Abingdon, Oxon OX14 4RN
605 Third Avenue, New York, NY 10017, USA

Library of Congress Cataloging-in-Publication Data

Anthony J.
smission line reliability and security/Anthony J. Pansini.
 p. cm.
J 978-0-8247-5671-0 (print) -- ISBN 978-8-7702-2244-0 (electronic)
 Electric power transmission. 2. Electric power systems--Secu-
ures. 3. Electric circuits--Reliability. I. Title.

091.P3132 2004
319--dc22

                                              2004043334

ion line reliability and security/Anthony J. Pansini.
ished by Fairmont Press in 2004.

is an imprint of the Taylor & Francis Group, an informa business

71-3 (The Fairmont Press, Inc.)
7-5671-0 (print)
2-2244-0 (online)
1-5115-9 (ebook master)

 effort is made to provide dependable information, the publisher, authors,
cannot be held responsible for any errors or omissions.

To the three women in my life...

My mother, My wife, My daughter

Contents

Preface

In 1992, Congress passed legislation essentially permitting electric (and other) utilities to be unregulated, resulting in transmission displacing local generation as the main supply to local communities and creating problems of service reliability. In 2001, events associated with 9-11 created problems of security. Although solutions to both can be coordinated, focus must be directed on security. Attention will be given to prevention, damage limitation and rapid restoration

Criteria for arriving at solutions are: Electrically, circuits must be so arranged that the shutting down of one or two does not cascade into an area blackout. Mechanically, new circuits should have a low profile so as not to attract undue attention; fences, surrounding the structures should be provided to thwart saboteurs from their intentions; circuits should be designed so that any damage or destruction is limited; and procedures updated for rapid restoration of facilities when several simultaneous points of damage or destruction are involved. For new and existing lines, structures should be surrounded by barbed wire topped fences and the installation of battery operated sensors that would radio back to the system operator an alarm of intrusion and its location.

The criteria may also limit the voltage class of the transmission line to 138 kV because of code clearance requirements of distances between conductor and ground and, between conductors, take up space available on the tallest wood structure. This also happens to be the limit for 'solid' type insulation (no gas or oil accessories) underground cables that may be installed in sections or branches of the circuit, and it is also the approximate limit for working from insulated bucket vehicles.

Meeting the criteria mentioned above, the usual economic considerations may have to be strained to new limits, and in some instances ignored. Some attempt should be made to separate charges to anti-terrorist protection activities from those associated with the improvement of service reliability; the former may perhaps be charged to the federal government, while the latter may not.

Deregulation resulted in the utility being restructured into three separate and independent enterprises: generation, transmission and distribution (although almost all three became parts of the original utility as a holding company). Allegedly, this mode of operation would result in competition between the several owners of these new operating units, allowing any single consumer the opportunity to choose his or her supplier, taking advantage of less expensive sources of energy—with the quality of service equal to or better than that of the replaced utility.

This discussion of the transmission systems would not be complete without some explanation of the effect on and by these three units on each other.

Increasing demands for electric service required the timely construction and installation of new generating facilities. For environmental and economic reasons, sites for such new installations were/are becoming extremely difficult, if not impossible, to find somewhere within the franchise area (not to mention load centers) of each of the several independent companies.

The situation was/is mitigated by allowing new construction to be postponed by the introduction of so-called cogeneration or merchant generation. Here, larger industrial and commercial consumers are encouraged to install their own generating facilities with excess generation during their cycles of operation to be sold to the local utility by connecting to their transmission or distribution systems (a reversal of policy in some instances). In some states, this consumer generation is not only mandated, but its unit cost is also set at that of the least efficient unit of the utility's generation sources.

Later, a smaller version, known as distributed generation, was developed to be connected to the distribution system at strategic points. These mainly consisted of small units, generally driven by small gas turbines, but also including a few fuel cell experimental applications; these may be both utility or consumer owned. These units, and the cogeneration units are not usually competitive with utility owned units that have the advantage of economy of scale.

Some cogeneration and distributed generation units may impact negatively on the safety of operations. Although standards for the selection, installation and maintenance of equipment to connect and disconnect these units from the system to which they supply electric energy are furnished the consumer by the utility, these standards are not always followed, particularly those relating to maintenance. This constitutes a hazard to persons who may be working on the system; believing it is de-energized, they may be the victims of an improper, unannounced connection energizing the system to which they are connected. Similarly, should a fault develop on the utility system to which they are connected and the equipment fail to disconnect their generation from the system, overloads and fires and explosions may occur. Further, while they are under the supervision of the system operator, they tend to dilute his attention to other events occurring on the electric system to which they may be connected.

Transmission systems, restructured into separate entities for privately owned and operated utilities, have assumed a more important and sensitive position in the supply chain of electricity to the ultimate consumer. Their role has essentially been reversed, from providing backup and peak power to local generator based systems. They now will become the primary source of supply, with local generators (if any) as backup and peak power sources for the new transmission systems.

In this respect, the restructured transmission systems essentially follow those of some publicly owned power systems, such as TVA, Hoover Dam, Grand Coulee, and others—systems designed and operated to supply large amounts of available power to load centers at relatively long distances from the power sources.

In both instances, the associated transmission lines are located generally away from populated centers in sparsely inhabited areas, in the "boondocks," for economic and environmental reasons. Often, the structures support two circuits, and occasionally more. They are very vulnerable, not only to vandalism and the vagaries of man and nature, but especially to the saboteur, as they are extremely exposed.

No attempt is made in this dissertation to describe the detailed planning, design, and normal operation and maintenance of transmission facilities. Many excellent works exist sufficient to fill the needs of such endeavors. Here, the discussion is based primarily on the effect of deregulation on the reliability of the newly deregulated systems, and consideration of what may be done to maintain or improve the reliability of the now deregulated transmission systems. Some phenomena occurring in electric theory are included, in non- or semi-technical terms, for the benefit of non-technical people and to refresh the memories of engineers.

It is obvious that reliability and security go hand in hand. But whereas reliability encompasses contingencies resulting from flaws in design and human errors, security tends to negate damage and destruction of property and injury or death to humans deliberately caused by other humans.

While steps are taken to provide security for generating stations, the transmission lines, which are the delivery systems for their product, under the deregulated system, provide literally thousands of miles of opportunity for saboteurs to deny that product from being delivered to consumers, all with relative ease and safety.

Electric service has become a necessity not only in the lives of individuals, but also in the operation of public services; e.g., water supply, sewage disposal, communications, transportation, health activities, etc. Already military surveillance and protection has been extended to nuclear power installations, and probably may be extended to large power generating facilities. As the main source of supply to large areas of the country, should not transmission facilities receive attention, possibly by some sort of paramilitary agencies? And/or, should not another look be taken at the methods of supply of electric energy, in view of events of September 11, 2001?

Chapter 1

Circuit Arrangements

The choice of the type of electric circuits in the transmission system as the primary source of supply is of great importance as it plays a great part in achieving high levels of reliability, including security, of electric service to the consumers.

Under deregulation of electric utilities, transmission systems, as the link between generation in the chain of supply between generation and consumers, will assume great importance. They will supplant local generation as the primary supply to consumers. As important as the transmission lines that will supply loads of diverse characteristics and sensitivity, the type of circuit selected is important. These will have an influence on the economics as well as the design, construction, operation and maintenance of projected facilities.

A simplified diagram illustrating the changes brought about by deregulation, and useful as a comparison with present regulated operations, is shown in Figure 1-1. Note the introduction of new "super" transmission lines attached to relatively large generation sources. The changed function of transmission is apparent, namely, a reversal of functions, from backup and peak supply to local generation to the primary source of supply, with local generation (if any) as backup and peak supply to the new transmission systems.

The elimination of some or all of the local generation units with the completion of the new transmission sources of supply, especially their demolition, in order to reduce their capitalization to

1

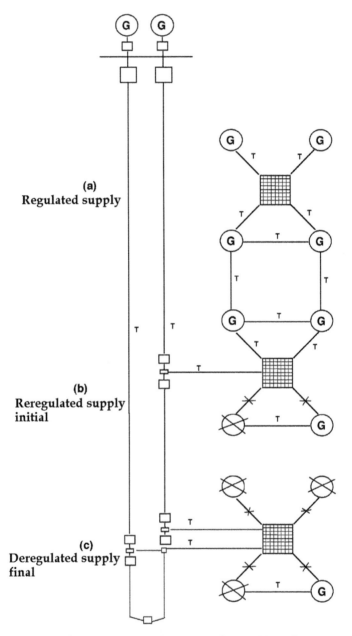

(a)
Regulated supply

(b)
Reregulated supply initial

(c)
Deregulated supply final

Figure 1-1. Simplified schematic diagram of transition from regulated to deregulated supply systems.

achieve competitive rates with other suppliers, should be unnecessary. The capitalization of such units can be reduced to one dollar for accounting, tax, and other legal purposes. If this violates some existing practices, steps should be taken to make them acceptable, at least for the duration of the security measures against terrorism in existence. Considering the vulnerability of transmission systems, located in sparsely populated areas for economic and environmental reasons (as new generation is planned to be) prudence dictates that these local generator units be retained during the period national security is challenged. Some billing adjustments to the consumer, similar to present fuel adjustments, during the period of their operation, can be considered.

The types of circuits that may be contemplated for the proposed transmission system, and their importance on reliability, are shown in Figures 1-2 for radial systems and 1-3 for loop systems. In Figure 1-2a, a fault on any part of this type circuit affects all of the consumers it serves, cutting off their supply.

In Figure 1-2b, service reliability may be slightly improved with the installation of sectionalizing switches, as shown. Here, a fault will de-energize the entire circuit until the sectionalizing switch between the fault and the station can be opened, possibly by automation; the circuit breaker at the station will reclose (if programmed to do so) and service restored to those consumers from the point of fault back to the station. The process is repeated after the fault is cleared; the entire circuit is de-energized by the station circuit breaker until the open sectionalizing switch is closed, then the station circuit breaker is closed, restoring service to the entire circuit.

In Figure 1-2c, the circuit arrangement is essentially the same as that of Figure 1-2b. The switches in that circuit are replaced with the much more expensive circuit breakers and

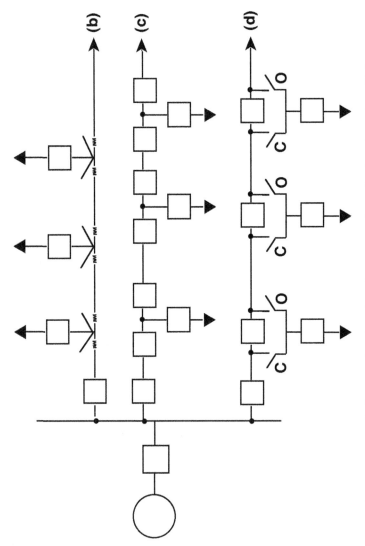

Figure 1-2. Radial type circuits showing several methods of sectionalizing.

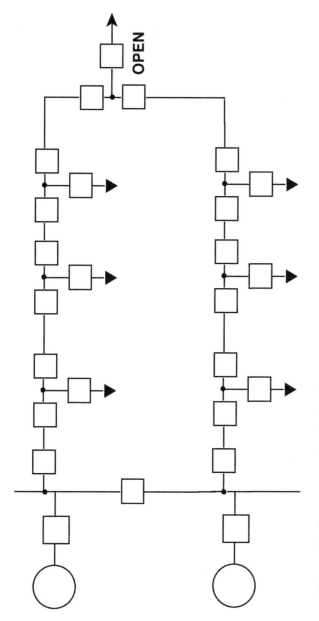

Figure 1-3. Loop type circuit showing methods of sectionalizing. One circuit breaker scheme as shown in Figure 1-2 also applies.

associated relaying. Since all the switching necessary to isolate the section on which the fault is located is done by circuit breakers, the intervals when the entire circuit is de-energized are shorter, and full restoration of service on the entire circuit is considerably shortened.

In Figure 1-2d, a less expensive arrangement, with fewer circuit breakers, is essentially the same as that in Figure 1-2c, except that the circuit breakers on both sides of the section on which the fault is located must be open before the isolating switches may be opened and also when closed.

In Figure 1-3, the circuit consists essentially of two of the kind shown in Figure 1-2c, connected together through a circuit breaker to form a continuous loop, capable of being sectionalized automatically between branches, de-energizing that section on which the fault is located. Faults on a branch or consumer are isolated by the operation of the circuit breaker connecting it to the main line. The arrangement in Figure 1-2d may replace the one described, with all the extra switching associated with that arrangement.

While this closed loop arrangement appears to be the best reliable one, there is danger that the fault may be so located that the fault current flowing in one branch may not be high enough to operate the isolating circuit breaker on that branch of the loop. By operating the loop in two sections, an open loop, total fault current will flow to the fault located between the open (loop) circuit breaker and the station, assuring the operation of the circuit breaker nearest the fault between the fault and the station. Meanwhile, the open circuit breaker, sensing one side is now de-energized, will close automatically, restoring the open loop and service to all the consumers on the circuit, except those connected to the isolated section on which the fault is located.

While the circuit arrangements described above are "basic" arrangements, combination of some of them may pro-

vide satisfactory and economic solutions to individual problems.

In Figure 1-3, two generators are shown connected together through a circuit breaker that would operate normally
open. Should it be found desirable to operate with it closed,
precautions should be taken to synchronize the two generators before closing the circuit breakers.

If in Figure 1-3, the two generators represent two separate generating stations distanced apart, then the loop represents a tie between the two generators. Should a fault now
develop, fault current would flow from each of the generators in proportion to the distance each is from the fault. This
could result in slowing down the unit supplying most of the
fault and load current, and a relative speeding of the unit
farthest from the fault. The result would be a continuing
rocking motion between the two generators, accelerating, so
that ultimately one would have its circuit breaker open from
overload, and subsequently the other would also have its
circuit breaker open, resulting in this part of the system to
shut down. The same effect would take place with a number
of generators similarly distanced and connected together.
The result would be that each generator would have its circuit breaker open, "cascading" one after the other, until the
entire system is shot down (blackout). To obviate this occur-
rence, it is imperative that the circuit breakers nearest the
fault open as fast as possible, clearing the fault from the circuit and stopping the rocking motion described above from
continuing. This activity is often referred to as the "stability"
of the system.

Protective relays play an important part in the operation
of a transmission system. They initiate the opening or closing of the breaker that may take a fraction of a second to
complete. The relay itself may take even less time to function, making the length of the circuit from the relay to the

mechanism operating the breaker a factor to be considered. Where the coordination of operation of several breakers is involved, the relaying circuitry may be complex, and hence prone to malfunctioning. In this case, redundant relaying should be considered. Where the breakers involved are distant from each other, communication between relays associated with them is necessary, and this may be accomplished by pilot wire, leased telephone wire, wireless, and, of recent usage, fiber optic conductors may be considered. Present-day electronic relays are much faster than the older electro-magnetic type and retrofitting of such older relays is recommended.

In planning the circuitry of transmission systems, the associated relaying required to accomplish the desired results must always be taken into account. Sometimes it may be the determining factor in the circuitry chosen.

The principle of the sectionalized open loop circuit is applicable to transmission lines, the moveable open point allows transfers of loads in both normal and emergency situations, limits service interruptions only to the section on which the fault occurs; the only time the loop is closed is momentary during switching times. The loop may originate and end at the same station, or may constitute an open tie between two stations. The opening in the circuit prevents disturbances (e.g., overloads, stability) from being communicated to other stations. Accomplishing this at stations by means of buses is detailed in Chapter 4, Substations.

Chapter 2

Electrical Criteria

Alternating current circuits, when energized, create electromagnetic and electrostatic fields about the conductors. These alternating (moving) fields in turn create alternating voltages within the conductors that do not coincide with the voltage supplied to the circuit. The net result is the current and voltage waves do not act in unison but become displaced with each other as shown in Figures 2-1a and 2-1b. The voltage generated in the conductor because of the electromagnetic field is 90 degrees behind relative to that of the main load current (Figure 2-1a), while that caused by the electrostatic field is 90 degrees ahead of that of the load current (Figure 2-1b). These fields do not exist separately in the conductor, but act together to displace the current and voltage waves in the circuit from each other; that is, they tend to impede the flow of current relative to the voltage. The resulting effect is that the power delivered now is less than the power supplied, and this ratio is known as the power factor of the circuit.

The effect of the voltage created by the electromagnetic field is called inductive reactance while that created by the electrostatic field is called capacitive reactance. The two reactances tend to balance each other; the net effect is generally termed the reactance of the circuit.

The actions described above also appear when two (or more) such energized conductors are adjacent to each other

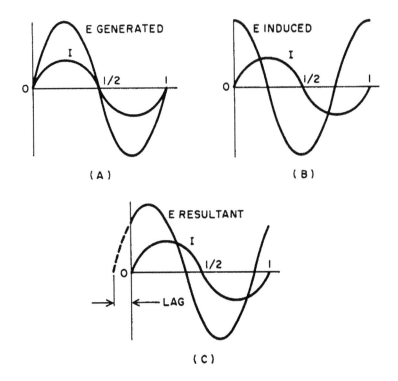

Figure 2-1a. Effect of inductance on voltage and current in a conductor (not to scale).

(Figure 2-2), further affecting the relationship between the current and voltage waves within the conductors. The effect of this relationship is referred to as the mutual reactance of the circuits.

The amount of these several reactances of a circuit depend on the length of the circuit as well as on the length of the adjacency of nearby conductors, and the values of the currents in these conductors. In the case of transmission lines, these quantities may be very significant, restricting severely the capacity of the line. There are ways of overcoming these restrictions.

The usual transmission circuit consists of three ener-

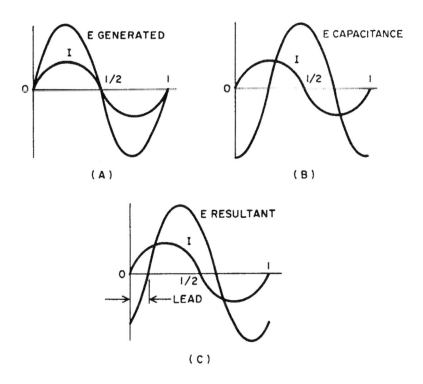

(A)

(B)

(C)

Figure 2-1b. Effect of capacitance on voltage and current in a conductor (not to scale).

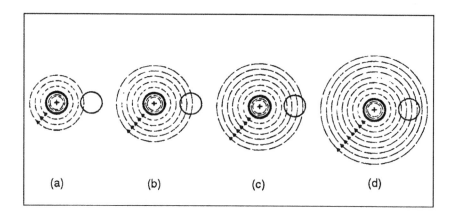

Figure 2-2. Effect of the magnetic field about a conductor on an adjacent conductor.

gized conductors and a neutral or ground wire (Figure 2-3). Each conductor carries approximately one-third of the power flowing in the circuit (exactly one-third when evenly balanced). Each conductor is known as a phase; the circuit, therefore, is a three-phase circuit. Each phase is electrically 120 degrees apart from each other. The "disturbing" reactances associated with the position of each phase are also displaced 120 degrees from each other. Now, if each phase conductor can be relocated in its position relative to the other two for one-third of the circuit length, for the entire circuit, they will tend to cancel each other. The one-third length need not be all in one section of the line, but may consist of several transpositions totaling approximately one third. This is done with three phase length lines.

For shorter lines or branch lines of only one or two phases, this may not prove practical. Here banks of inductive

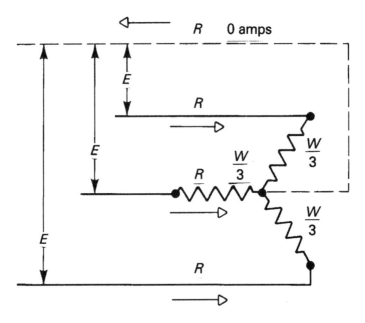

Figure 2-3.

reactors and/or capacitive reactors are installed on such lines (Figure 2-4a), to counter the reactance of the line, helping the line to achieve a 100 percent power factor, or as close to this as practical. As the load varies, the transmission line may vary daily, weekly, monthly, etc.; the units of the reactors are equipped with circuit breakers that can control the amounts of corrective inductive or capacitive reactance required to achieve the goal (Figures 2-4b and 2-4c).

Combinations of both methods may be employed where it may be the solution in some instances.

One other distinguishing feature of alternating current circuits is known as the "skin effect." The density of the current flowing in a conductor, particularly at the higher

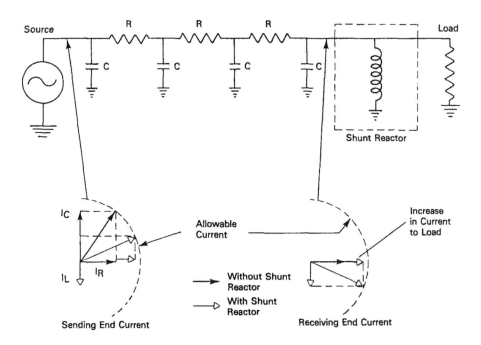

Figure 2-4a. Simplified cable circuit (above) shows how power delivered to load is affected by a shunt reactor. In the absence of shunt reactors, this power decreases to zero at some critical length.

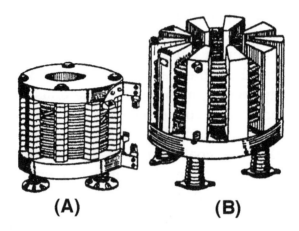

(A)   **(B)**

Figure 2-4b. Current-limiting reactors: (A) Westinghouse, (B) General Electric.

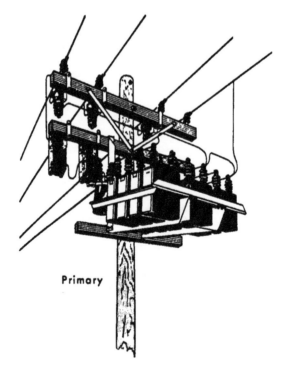

Primary

Figure 2-4c. Capacitor installation on poles.

operating voltages, tends to be greater toward the surface of the conductor and reducing as it nears the center of the conductor. This results in the use of hollow type conductors descried in the next chapter.

LIGHTNING PROTECTION

Strictly speaking, this is not an electric circuit consideration, but does affect its operation.

Generally, the object of such protection is to prevent the atmosphere surrounding the transmission lines from becoming highly charged allowing a relatively low resistance path to form for the discharge of the electric charges forming on a cloud when the voltage generated there reaches a critical point. This is achieved in two ways: the use of an overhead ground or shield wire, and of lightning arresters.

As the name implies, the shield wire is placed over the conductors at a height to form a 30-degree angle (from experience and tests); however, it must not be inordinately high as its effect diminishes the higher away from the conductors (Figures 2-5a and 2-5b). In that case, two (or more) such wires need to be installed. These wires are solidly grounded at each structure where they are supported. This connection includes all other points of ground on the structure, including the ground of the circuit or circuits themselves.

The lighting arrester (Figures 2-6a and 2-6b) is installed on the structure supporting the top conductor of a circuit in such a manner as to have one end (or a short conductor from the arrester) within a determined space. When the voltage created by the charged atmosphere and the circuit voltage become great enough, a discharge across the air gap occurs, creating an arc that is made to enter a tube where the pressure of the confined air tends to blow out the arc, restoring

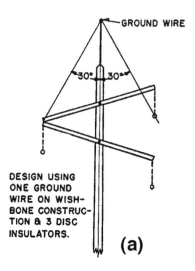

DESIGN USING
ONE GROUND
WIRE ON WISH-
BONE CONSTRUC-
TION & 3 DISC
INSULATORS.

(a)

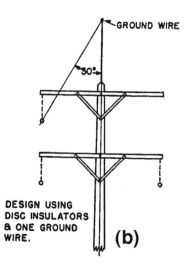

DESIGN USING
DISC INSULATORS
& ONE GROUND
WIRE.

(b)

Figure 2-5. Showing 30° angle for lightning protection on overhead wires.

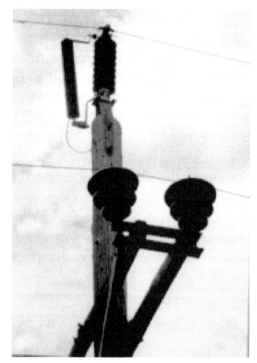

Figure 2-6. 69 kV Lighting arrester.

the arrester to its protective position. It is highly important that the capacity of the arrester be large enough not to have the tube fail, causing the arc to be diverted across the insulators (that may be wet and/or dirty) to the structure, where the arc may damage both the insulators and the structure seeking a path to ground. The flashover of the insulators constitutes a fault on the line and may trip the circuit breaker back at the source station. Although only one arrester may be installed on the topmost conductor, there are occasions where arresters on all conductors may be desirable. Usually arresters are installed in addition to the existence of the ground or shield wire.

To insure a good ground, copper rods are driven into the ground at the foot of the supporting structures and are interconnected with other grounds. Sometimes, a grid or mesh of bare copper wires is buried at the foot of the structures and connected to other grounds; this is referred to as a "counterpoise" (Figure 2-7, see also Figure 3-7c). Additional grounding may be obtained by connecting these counterpoises with a buried bare wire. All of these increase the area for the current flowing to ground to be dissipated harmlessly into the ground.

The importance of good grounds is emphasized, not only for lighting purposes, but also for proper operation of protective relays, and especially for safety reasons.

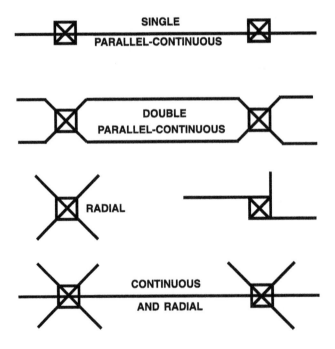

Figure 2-7. Arrangements of counterpoises.

Chapter 3

Mechanical Criteria

There are obviously many items in this category from rights-of-way to, but not including, the substation; substations will be the subject of Chapter 4.

The right of way over which a transmission line is located depends largely on the type of line and the nature of the terrain. In general, it should be wide enough so that adjacent trees or other structures, in falling, do not encumber the space under the conductors; conversely, should the conductors fall, they do not encounter trees or other structures at which they may cause fire (Figure 3-1). The area of the right of way should be kept clear of vegetation that would

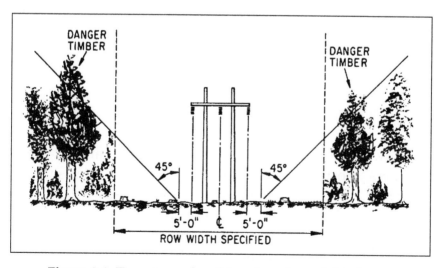

Figure 3-1. Tree removal and topping along right-of-way.

not only impede the transportation of maintenance and repair crews, but also constitute a fire hazard.

Access roads to the rights of way should be unobtrusive
and located at strategic locations to ease the work of crews
maintaining the right of way as well as the transmission
lines. If gates are involved, they should be locked with keys
or coded strips available only to restricted personnel. Where
they are in areas subject to vandalism, fences may be erected
for a few hundred or thousand feet on either side of the
access point, with rolls of barbed wire at the top and bottom
of the fence (Figure 3-2). In extreme cases, the entrance may
be electrified. In any case, ample notices should be posted to
keep out people who have no business there.

Where helicopters are used to patrol the lines or for
carrying men and material for repairs, landing pads should
be provided, possibly near the access locations. In areas difficult to access, either extra conductors may be installed or
caches of material shored in a weather-proof fashion.

All of these precautions, and others, may not discourage
the saboteur from attempting to damage or destroy the overhead line, but would make his task more difficult and perhaps limit the extent of the sabotage.

STRUCTURES—TOWERS

Towers are generally fabricated of galvanized steel
members, mostly of angle iron but including other shapes,
bolted or riveted together, and in some cases welded together. Their height may vary from some 50 feet to 150 feet,
depending on the voltage of the transmission line (Figure 3-
3) and the minimum standards of distances of the lowest
conductor at the lowest point of sag to the ground and clearances between conductors, in accordance with the National

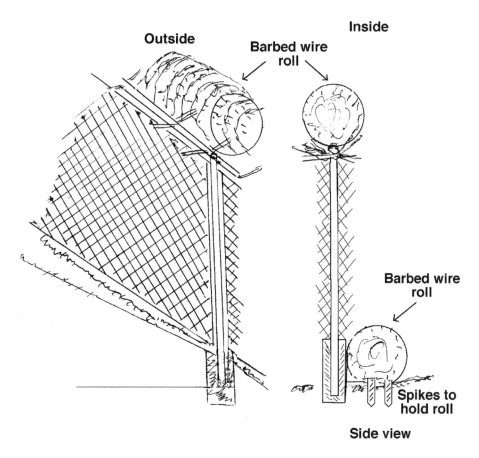

Outside

Inside

Barbed wire roll

Barbed wire roll

Spikes to hold roll

Side view

Figure 3-2. Barbed wire fence.

Electric Safety Code. The framework nature of the structure minimizes the effect of wind and generally provides a means of access for the worker, although steps are sometimes provided. The tower footings are attached to sister members imbedded in concrete pilings for short, light towers and in concrete foundations for the larger, heavier duty towers. Multiple bolts or rivets are used, again depending on the forces they must sustain. There are several types of towers, each designed for its purpose.

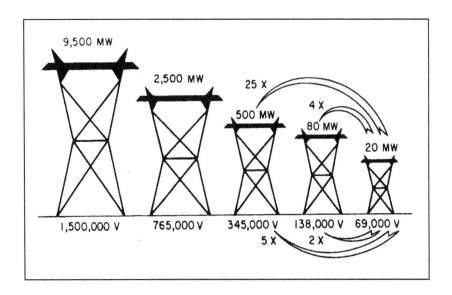

Figure 3-3. Comparison of power-carrying capabilities and operating voltages.

Suspension or Tangent

Suspension or tangent type are towers in which the insulators to which the conductors are attached are free to swing (Figure 3-4). The spans on either side of the insulator string approximately even so that the structure carries only the weight of the conductors, ice coated, and the force of the wind against them.

Angle or Corner

Angle or corner type towers (Figure 3-5), as the description implies, are designed to hold the spans of conductors on either side that are at an angle with each other. They are usually also of dead end design. These towers are of great strength, and may also be guyed to take care of the forces due to the uneven balance of conductors attached to the crossarm on the tower.

GUYS

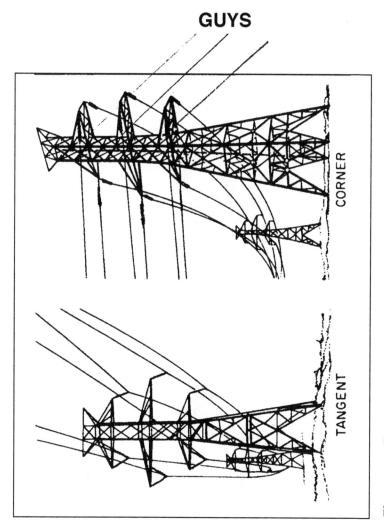

CORNER

Figure 3-5. Angle or corner tower.

TANGENT

Figure 3-4. Tangent or suspension tower constructions.

Dead-end

Dead-end type towers (Figure 3-6) also carry the weight of the span of conductor, but the spans may be unequal in length and the spans on either side are dead-ended, each mounted through the insulators directly to the crossarm attached to the tower. The strength of this type is greater than that of the suspension tower, designed purposely to withstand failure if all the conductors on one side break and no longer balance the ice-laden, wind-driven conductors on the

Figure 3-6. Dead-end tower.

other side. The towers are so designed to limit the damage, should the conductors and intervening suspension towers fail.

The great advantage of the tower (as compared to wood pole structures) is that it permits long spans of construction minimizing the number of such structures. The great disadvantage lies in the extent of damage caused by the failure of as little as one conductor or as great as the failure of a tower. Repairs are of usually long duration. In most instances, temporary wood pole lines are constructed around the failed portion to return the line to service as quickly as possible.

Wind blowing against ice-laden conductors, attached to the tower at a height, creates a moment arm that strains the tower legs, footings and connecting bolts or rivets. Moreover, the swaying conductors tend to weaken the mechanical strength of the conductors at the points of attachment. In some instances, the wind blowing at the conductor may cause near horizontal ice formations on a side of the conductor that, acting as an airplane wing, cause the conductor to rise and fall with the gusts of wind, causing the conductor to "dance," creating further strains on the tower structure.

Needless to say, the metal tower structures must be electrically grounded to take care of any flashover to the structure that may occur, leading the flashover current to ground safely.

Barbed wire strands are sometimes installed at the lower part of the tower to keep intruders from mounting the structure. In addition to this, since 9-11, it may be prudent to build a barbed wire top and bottom fence (see Figure 3-2), but with no gates (access by bucket vehicle) around the base of every tower. A coordinated attempt to blow up one or more towers on a major transmission line or on several transmission lines, can seriously disrupt the functioning of industrial plants (also defense plants) as well as the lives of populations in large areas of the country from long periods without electric service.

WOOD POLE STRUCTURES

Wood structures made of poles and timbers are generally confined to lower voltage transmission lines for economic reasons, but are also easier and quicker to repair. The wood members are generally of native timber (e.g., western cedar in the western regions, long leaf yellow pine in the South and East), chemically treated to resist rot and insects. Maximum pole lengths are in the nature of 70 feet, which limits the space available on them for carrying conductors and observing code clearances. Their strength also limits the loads (forces) they can sustain with acceptable margins of safety. Hence, span lengths are generally limited to about 500 feet, some one-quarter those of steel tower lines. Poles employed in the structures, as well as those employed singly on some transmission lines, are usually buried directly in the earth for their support. The same design considerations (ice, wind, lightning, code clearances, etc.) apply as for tower lines, although wood provides some extra insulation value.

In general, the same rules apply to wood pole structures, clearances, grounds, dead-ends, lightning arresters, etc. As wood is an insulator, wires connecting the overhead ground wire and other items on the pole are run down the side of the pole and connected to ground rods and/or counterpoises. Barbed wire fences should surround each pole structure, similar to those around towers.

Various configurations of steel towers and wood pole line structures are shown in Figures 3-7a, 3-7b, and 3-7c. Spans on wood pole lines are much shorter, generally not exceeding 500 feet, with the sag in the conductor spans usu-

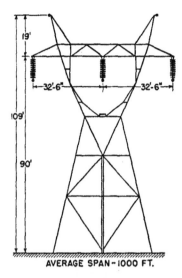

Figure 3-7a. Typical configurations of steel-tower lines.

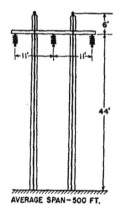

Figure 3-7b. Typical configurations of wood-pole lines.

ally so small that the danger of them contacting each other in the wind, or of "dancing" is not present. Failure of conductors or poles from ice and wind is more rapidly repaired, than for towers (mentioned above) because maintenance work, even at mid-span, both while energized or de-energized, can be done from insulated bucket vehicles (perhaps

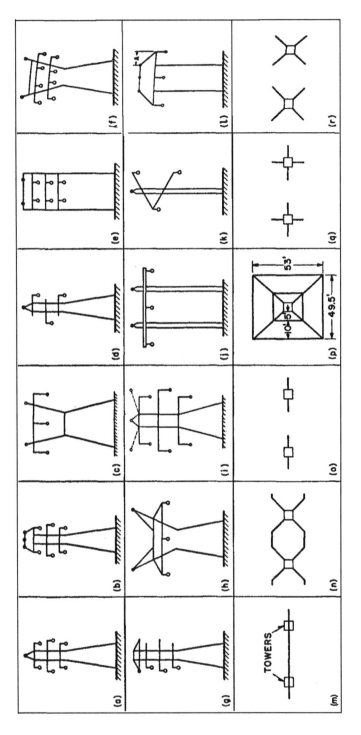

Figure 3-7c. Typical transmission structure and counterpoise configurations.

the greatest advantage), Figure 3-8). However, the greater number of structures (compared to towers) means that much more exposure to accidents and sabotage. And the same general protective fences and barbed wire installations apply.

Where the exposure to damage and destruction for both towers and wood pole structures may be unacceptable, recourse may be had to rerouting the portions of the overhead lines in peril, or placing them underground (Figure 3-9).

INSULATORS

Line insulators are the medium between energized conductors and their non-energized supports, points of attachment to the structure. They come in two popular shapes, discs and posts. Both consist of units that are capable of being added in strings for discs and in columns or posts to achieve the desired amount of insulation. They are made of porcelain and polymers (plastics), Figure 3-10. Other insulators may be found as supports for buses, on disconnecting switches, and as bushings on circuit breakers, transformers, and other devices. All are subject to accumulations of dirt (and salt in seaside areas), and are subject to cleansing during rain storms. All, however, are also subject to accumulation of ice and snow that tends to shorten the insulating path of the insulator and aid in the flashover of the insulator from nearby lightning strokes and switching surges.

The effects of such flashovers are more visible on the porcelain insulators than on the plastic ones and the cleansing effect of rain on the plastic insulators is not as effective as on the porcelain ones. The problem of external pollution and moisture has been recognized, and special vulcanized rubber coatings or the use of special rubber shields to in-

Figure 3-8. Insulated bucket vehicle.

Figure 3-9. Underground construction.

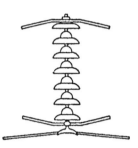

Figure 3-10a. Disc insulator with arcing rings or horns to keep flash away from insulators. (From Overhead Systems Reference Book).

Figure 3-10b. Typical post-type porcelain insulator. (Courtesy Ohio Brass Co.)

crease the surface creeping distance, the shortest distance measured along the surface of the insulator, have been developed. In other cases additional insulators have been added to the string. Caution should be taken in live wire (hot stick) work on the plastic ones; if in doubt, de-energize the work area. Safety should always be the first consideration.

Insulators to which conductors are attached are special targets for vandals and saboteurs as they can be shot at from a distance, often undetected. Damage or destruction of several discs in a chain, or post supports can cause flashover to ground that may cause circuit breakers to trip, de-energizing the circuit. Those on wood pole structures are perhaps more vulnerable than those on metallic towers, but the insulating factor of wood does tend to make flashovers more difficult, less frequent.

LIGHTNING ARRESTERS

Lightning arresters are installed at strategic locations on the transmission lines, usually some three or four spans apart. Their ground wires are connected to the overhead ground wire and to the steel structure on tower lines and to copper wires alongside the wooden structures to grounds in the earth or counterpoises where they exist.

CONDUCTORS

Conductors on transmission lines are usually made of copper or aluminum, and each circuit will consist of three conductors depending on the load to be carried and the allowable drop in voltage or electrical pressure. The conductors may be stranded alone for lower voltage transmission lines and stranded hollow core wires for the higher voltage lines (Figure 3-11). They may be filled with strands of inert material to hold the metallic strands in place along the outer part of the circle to take advantage of the "skin effect" of alternating current. For long spans, the core of the conductor may contain a steel wire or cable for mechanical strength

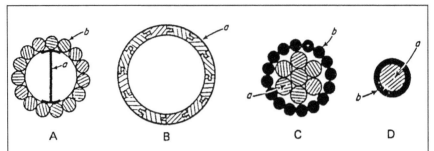

A   B   C   D

A   Hollow copper conductor with an I-beam core. *a*, I beam twisted spirally in a lay opposite to lay of individual wires *b*.

B   Type HH conductor using curved segments *a* laid in spiral configuration. Individual segments slide back and forth slightly when conductor is wound on reel.

C   Aluminum conductor steel reinforced commonly called acsr. *a*, steel strands making up the core; *b*, current-carrying aluminum strands.

D   One form of copperweld conductor. *a*, round steel core over which thin copper cylinder *b* in welded.

and may also contain some filler material to keep the steel wires centered in the conductor. The conductors are attached to the insulators holding them by means of "shoes" that are placed on the top and bottom of the conductor to spread the pressure applied in holding them there, this to avoid hot spots if a narrow clamp were used and the pressure applied at only a very narrow place on the conductor. Care should also be taken not to break any of the strands of the conductor as this not only may reduce the current capacity of the conductor, but also create a hot spot in the conductor that may cause its ultimate failure.

Where the structures (towers or wood poles) are able to carry additional loads, the capacity of the line may be doubled by installing a second conductor alongside the original one by means of hardware designed for such purposes (Figure 3-12).

Overhead ground wires may be of steel, copper, or copper clad steel, and wire connecting them and other metallic details are usually made of copper.

CABLE

Branch circuits (from the main transmission line) that supply large consumers as well as communities are more exposed to the would-be saboteur. They are more easily accessible because of the usually lower structures and, while still in less populated areas, closer to traveled roads. They are of usually lesser voltage than the main transmission line, generally at 138 kV or lower. This makes them candidates for underground cables, particularly as cables of these ratings are now available with "solid" type insulation requiring no oil or gas accessories. Although obviously more expensive and faults are more difficult to find and repair, they offer the best defense against sabotage.

The obvious economic advantage of overhead construction for transmission, and especially of tower lines, will continue, but may give way to wood pole lines during the period of possible terrorist acts of sabotage, and to underground cables in particular instances.

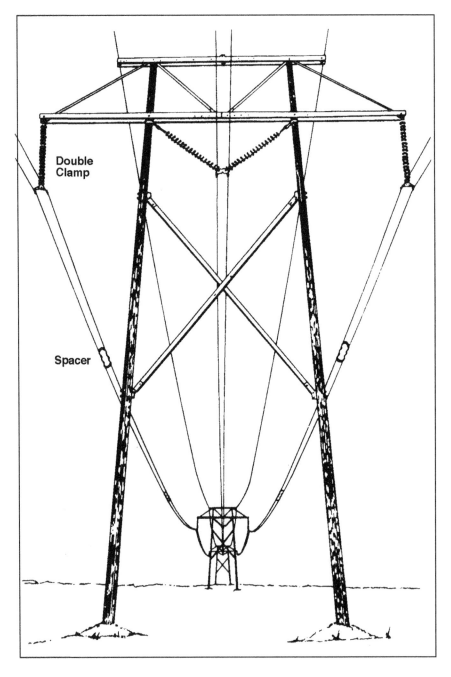

Figure 3-12. Double conductor line.

Chapter 4

Substation Criteria

Transmission substations are key points in transmission systems, especially in contingencies, as this is where are located circuit breakers and the relaying that controls them, transformers that (usually) reduce voltages to lower level branch circuits, metering to record operations, and possibly also banks of reactors and capacitors where they are needed for power factor correction. Also located here are accessories for all this equipment. Substations may also serve as headquarters for maintenance and restoration activities in event of emergencies.

Incoming and outgoing circuits are connected to buses through circuit breakers (Figure 4-1). The buses are so arranged that circuits may be sectionalized here to isolate a fault on any circuit, bus, transformer and the buses themselves. Some are done automatically with circuit breakers; others by means of isolating switches that must be de-energized before and after operating them—with temporary outages occurring in the process. Breakers here may also act as the open point in loop circuits, aiding in the sectionalizing of the circuit to isolate a faulted section (refer to Chapter 1). Note the flexibility of such operations increases with the addition of circuit breakers and the arrangement of buses. If several circuits converge into a station, the bus arrangement is such that loads between circuits may be redistributed or circuits energized that may not be faulted.

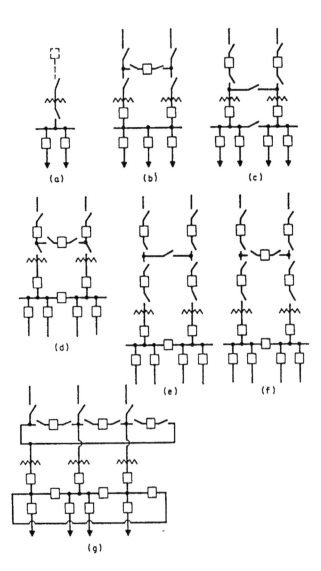

Figure 4-1. Incoming and outgoing feeder circuit breaker arrangements (*Courtesy Westinghouse Electric Co.*)

Where branch circuits are involved, the main circuit may supply transformers to lower voltage to the required level (Figure 4-2). Here again, the output of the transformers is connected to buses for similar flexibility as those described above. For service reliability, three circuits are recommended

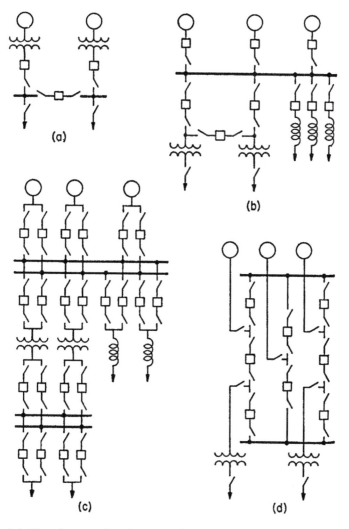

Figure 4-2. Fundamental schemes of supply at higher than generated voltage.

for the branch supply, two from one main source and the third from a separate source, if available, left open if it is desired to operate as an open loop circuit. Refer to Chapter 1.

Reliability of both incoming and outgoing circuits is provided by the bus arrangement. From least to some high degree, they all depend on the use of and arrangement of circuit breakers. Each additional circuit breaker provides greater flexibility of operation with a greater degree of reliability, including the ability to sectionalize both incoming and outgoing open-loop circuits at that point. The greater the number of circuit breakers, the greater the cost. Figure, 4-2 shows the application of these arrangements when the, incoming circuit voltage is to be raised to a higher outgoing voltage circuit. Note the application of reactors to limit the supply, of fault current to generators where transformers do not intervene in the circuitry. Figure 4-3 shows a modifica-

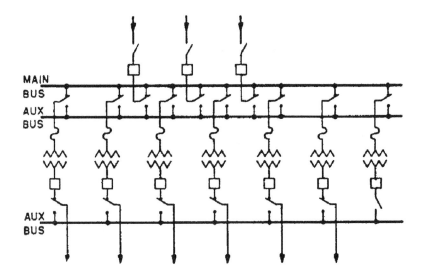

Figure 4-3. Distribution substation with high-voltage double bus and low-voltage auxiliary bus and individual transformation in each primary feeder resulting in relatively low interrupting duty on feeder breakers.

tion of the ring bus of Figure 4-1g where individual transformers are associated with each outgoing circuit, lessening the strain on their circuit breakers. Combinations of all of these bus arrangements often produce satisfactory reliability results at considerably reduced cost.

The bus arrangement in Figure 4-4 is designed so that each incoming circuit supplies a fixed number of outgoing circuits, each independent of the other. Here, a fault on one outgoing circuit does not interrupt service on the others, and likewise, a fault on one of the incoming circuits interrupts service on only the outgoing circuits it supplies. Nowhere are the incoming circuits connected together in a grid, nor are the outgoing circuits connected in a grid. Here, interruption of supply from any one circuit will not communicate to another on both the incoming and outgoing sides; there will be no possible overloading of one circuit from attempting to pick up the load supplied by the interrupted circuit, and hence, no cascading into total blackout. Obviously, there will be interruption of service to a given area, the load of which may or may not be picked up by other circuits at the command of the system operator, depending on the available spare capacity of the non-interrupted circuits that may be variable depending on the time of day. While, reliability is thus affected, it limits the operations of the saboteur—the price in reliability paid for a higher degree of security.

Buses should be physically separated a sufficient distance so that failure of one, with possible attendant explosion and fire, does not communicate to the other buses or to any other vital equipment. Similarly, circuit breakers and transformers should follow the same separation principle, achieved with steel reinforced explosion and fire proof barriers between them (that may also act as sound barriers Figure 4-5), and sump pits dug beneath each of these units sufficient to contain the oil in that unit even if aflame.

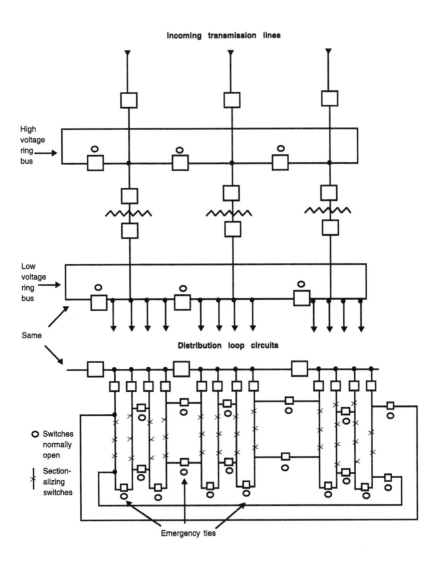

Figure 4-4. Substation bus arrangement.

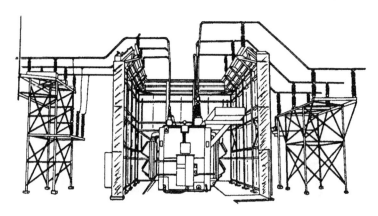

Figure 4-5. Substation power transformer with barriers.

It was previously recommended that outgoing circuits from the substation be underground cable. If, for some reason this should not be possible or practical, it is suggested that at least a significant distance for the first part be underground to confuse saboteurs.

Lightning protection should follow the same principles as for overhead lines, described above. The area of such substations will be quite large and its perimeters should be contained by barbed wire fencing, and in extreme cases, electrified, all as mentioned previously . Consideration should be given to erecting a wire mesh over the substation, or parts of it, to prevent sabotage from the air; this would also serve as lightning protection.

Possible underground construction of such substations should be given consideration, as well as the transmission lines themselves, or at least portions of them in high sabotage areas. The economics of such construction are so staggering that other means of power delivery will no doubt be chosen.

All equipment in the substation should be connected through air switches for safety reasons (Figure 4-6). Whether

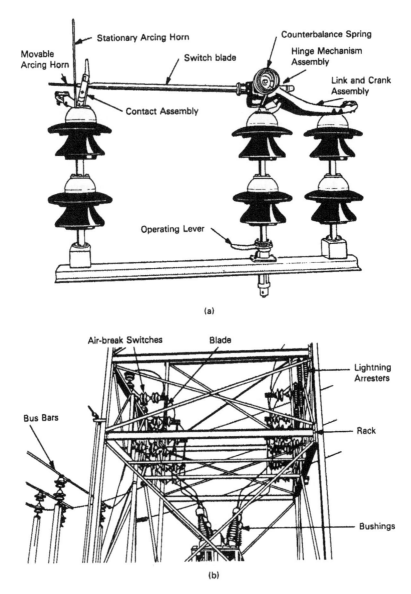

Figure 4-6. Air-break switches mounted on a substation rack.

for routine maintenance or in emergencies, the worker must be able to see an opening in the circuit on both sides of the equipment on which he, or she, may be working.

Substation buses are convenient places on which to install fault locating devices. These may hasten the location and elimination of faults. One suggested device is shown in Figure 4-7 (page 46). Although many are not 100 percent accurate; they are sufficient to give an approximate location, which together with other field indications, can speed the process of restoration.

Portable substations, Figure 4-8, should be kept on hand to replace temporarily units that may be so damaged that they may not be repaired in the field.

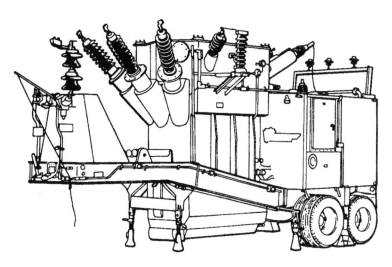

Figure 4-8. Portable transformer.

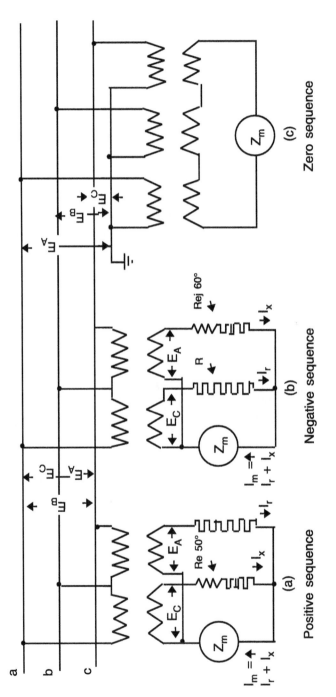

Figure 4-7. Circuit for sequence fault locator.

Chapter 5

Direct Current Transmission

There are relatively few direct current transmission lines in operation, mainly because of the high cost of transforming alternating current into direct current at one end of the line and then back again to alternating current at the other end, in a process that allows this to happen when current is flowing in either direction. However, it has so many other advantages, compared to alternating current, that it becomes affordable in certain instances.

One of its chief advantages is that one conductor may take the place of three in a-c conductors in a circuit. The return path for such a d-c circuit can be the ground, aided by the grounds associated with the lightning protection of the line, namely, the overhead ground wire and the underground counterpoises, some of which may involve a continuous path between structures (Figure 5-1). Crossing bodies of water, particularly sea water, the sea provides the return conductor.

Other important advantages stem from the fact that d-c circuits do not have alternating magnetic and electrostatic fields about them, hence no problems with inductive and capacitive reactances with their effects on voltage and power losses that may seriously reduce the active power transmission capability, requiring the use of corrective reactors. For the same voltage rating as a comparative a-c circuit, the d-c circuit requires only some 70 percent of the insulation (Figure 5-2) or, expressed differently, the same insulation as the a-c circuit can accommodate a voltage some 30 percent higher

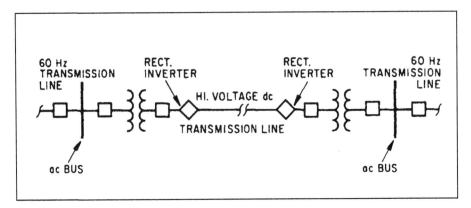

Figure 5-1. DC transmission used over very long distances. The two AC buses may be hundreds of miles apart and do not have to be synchronized, or in phase, to permit power to flow between systems.

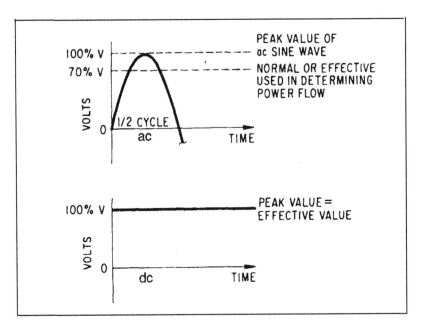

Figure 5-2. AC vs. DC power transmission. In AC, the peak voltage must be used in calculating the insulation required from conductor to grounded supporting structure. This value is higher than the effective value of the DC system shown at the bottom. The DC system utilizes the maximum voltage to ground to transmit power.

with an increase of the capability of the d-c circuit to deliver some 30 percent more power.

In connecting two d-c circuits of the same voltage rating, there is no need for synchronization; in similar a c situations, where the a-c circuits are of different frequencies (e.g. 50 and 60 cycle), conversion to d-c makes their connection feasible. All of this makes the control of d-c circuits simpler than their a-c counterparts.

Care, however, should be taken in energizing and de-energizing d-c circuits, as the rise and collapse of their associated magnetic fields may induce unwanted voltages in the conductor itself and in surrounding conductors. An automatic placing of temporary grounds on the switches during this part of their operation is usually provided to insure the safety of the workers. Care should also be taken not to have the d-c circuit in close proximity to an energized a-c circuit, as the a-c voltages induced in the d-c circuit may have some influence on the d-c "wave" and the existence of the a-c voltage will produce circulating currents that will only serve to heat the conductor and reduce the circuit capability; a-c filters are available that drain these unwanted circulating currents to ground.

A simplified schematic diagram of a typical d-c transmission line is shown in Figure 5-3. The a-c to d-c voltage rectifiers are usually thyrite units that are capable of rectifying a-c to d-c and, as inverters, converting the d-c back to a-c. The losses involved in their operation are relatively low and their maintenance (by replacing of worn out units) is also relatively low. Figure 5-4 shows a typical waveform that is an example of change from maximum positive a-c voltage to d-c voltages. Some idea of the size of the banks of thyrite rectifier-inverters for a 345 kV circuit is shown in Figure 5-5. Clearly, what is needed is a d-c transformer that approximates the simplicity and efficiency of the a-c transformer.

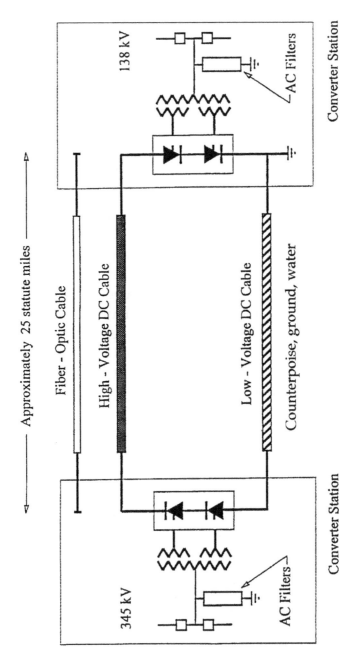

Figure 5-3. Simplified schematic diagram of a high voltage DC transmission line.

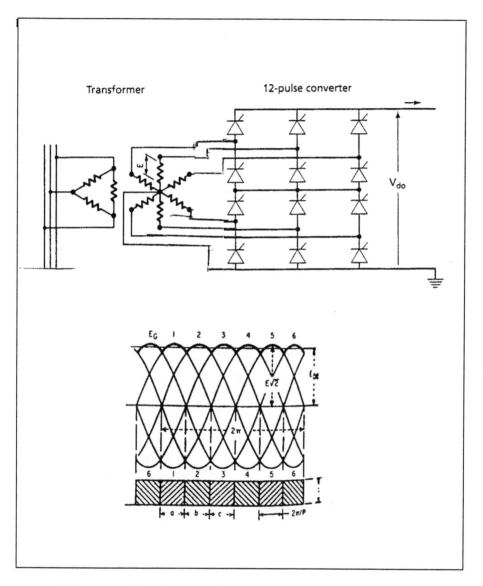

Figure 5-4. Six-phase rectification with twelve half circles.

Figure 5-5. Stack of thyrite rectifiers for 345 kV DC transmission line.

Chapter 6

Operations Criteria

ELECTRICAL

The transmission system should be such that the de-energization of one or more circuits should not result in cascading to a system blackout. This can be accomplished by not connecting them in a network or mesh, but having each circuit consist of an open loop configuration with sectionalizing facilities at each consumer branch, or substation, Figure 6-1. Moreover, each substation should have three sources, each independent of the others and acting as a sectionalizing point, see Figure 4-4. In this manner, the saboteur who may succeed in causing a failure simultaneously on one or more circuits will not achieve an area blackout, but only a few sections of circuits on which the faults may have been placed. The normal open in the circuit now transferred to the de-energized sections containing the faults. Emergency ties to adjacent circuits may be provided for those extremely rare cases where two or more sections of the circuit may be faulted (Figure 6-2). Relaying for such an arrangement may be difficult and perhaps not reliable, in which case an electronic scenario may be designed for such special cases.

This arrangement contains a relatively large number of expensive circuit breakers. Where short time interruptions of service can be tolerated while manual switching of disconnects takes place, the circuits may be modified, as shown in Chapter 1.

This arrangement may not result in the economy associated with the network or mesh arrangement, but this may
be a small price to pay for the greater reliability against mass
blackouts.

In almost all arrangements of circuits, in times of emergency when one or more circuits are called upon to pick up
some of the load of adjacent circuits, prudence would dictate
that, under normal conditions, circuits be not operated at or
near full load, but some comfortable margin be left to accommodate picking up the additional load of an adjacent circuit
in trouble.

MECHANICAL

Daily helicopter patrols, weather permitting, should be
made with bolometers scanning the circuits for the existence
of hot spots that may ultimately cause failure. A monthly
ground patrol to check for conditions not seen by air may be
desirable. These may include any visible attempts at forced
entry, damage to fences and gates, and the changing of batteries in the sensing devices. Insulated bucket vehicles (Figure 6-3) should be employed with their crews trained in live
line (hot stick) maintenance (Figure 6-4). Other normal maintenance, such as brush and vegetation clearing, checking of
lightning arresters, warning signs for dirt on insulators, etc.,
should be carried out.

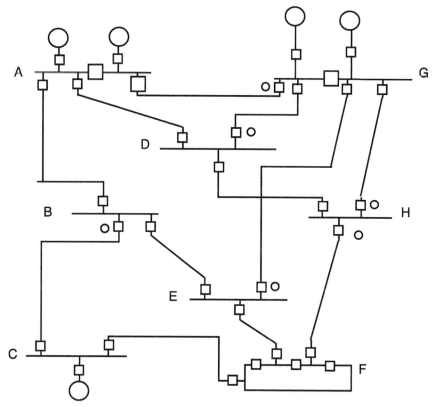

Sectionalizing of circuits not shown. Each station ring bus same as F.
Open circuit breakers acting as sectionalizing points.

Figure 6-1.. Open loop supply of transmission system (no grid or mesh)

Chapter 7

Economics

Evaluating the proposals included in the previous chapters is difficult in view of the different practices now employed by the several utilities. Some general remarks concern the rather obvious features; for example, the replacement of a mesh with open loop circuits. Specific features, such as the sectionalizing of circuits at each tap for consumers or branch lines with the arrangement of circuit breakers (and protective relaying). Obviously, there are areas where some interruptions can be tolerated for short or longer periods of time, as in some residential areas, while interruptions of any magnitude should be avoided in highly sensitive areas, such as water and sewage works, hospitals, communication centers, military installations, etc. In between these extremes are areas where short interruptions may be tolerated, as in public offices and schools, comparatively small industrial and commercial enterprises.

Knowing the time and cost for restoration of service, a comparison can be made of typical cases shown, with more expensive circuit breakers eliminated, or branch underground cables replaced with overhead lines, etc. And the dollars invested per number of consumers affected can then be compared to loss of revenue for the utility AND losses to the businesses involved. From these data, priority lists may be developed to act as guides in paring back features from circuits or parts of circuits to reduce investment costs. These indices may also be used in determining schedules for restoration in major contingencies.

In the final analysis, all of these factors should be tempered by local experiences to determine the final choice of circuit design, operation and maintenance.

MISCELLANEOUS

A few thoughts on research that should be expedited, perhaps by one new "Manhattan Project."

1. Conductors of zero resistance (or even negative values) under load. Work on this has been very successful, but much remains for it to be practical and economically competitive.

2. Development of a DC transformer that can rival its AC present counterpart.

3. Development of a means to store electricity directly, probably a DC solution.

4. A direct means of generating electricity. Present steam turbo-generators are about 30% efficient (from the fuel to the end product) and nuclear units are even much less.

ACKNOWLEDGMENTS

To Messrs. Kenneth Smalling and William Underwood for their review and comments on the material presented. To Ken Smalling for securing some of the data needed in the development of this work.

To my patient wife for putting up with my endeavors.

To the members of The Fairmont Press for their encouragement and their substantial help in the preparation of this work. Any errors, however, must be laid at my doorstep.

Appendix A

Insulation: Porcelain Vs. Polymer

For many years, porcelain for insulation purposes on lines and equipment has exercised a virtual monopoly. It was perhaps inevitable that plastics, successful as insulation for conductors since the early 1950's, should supplant porcelain as insulation for other applications in the electric power field.

The positive properties of porcelain are chiefly its high insulation value and its great strength under compression. Its negative features are its weight (low strength to weight ratio) and its tendency to fragmentation under stress. Much of the strength of a porcelain insulator is consumed in supporting its own weight. Figure A-1a&b.

In contrast, the so-called polymer not only has equally high insulation value, but acceptable strength under both compression and tension. It also has better water and sleet shedding properties, hence handles contamination more effectively, and is less prone to damage or destruction from vandalism. It is very much lighter in weight than porcelain (better strength to weight ratio), therefore more easily handled. Figure A-2, Table A-1.

Economically, costs of porcelain and polymer materials are very competitive, but the handling factors very much favor the polymer.

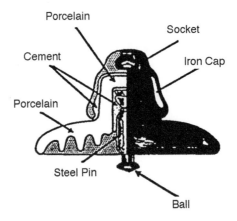

**Figure A-1 a. Ball and socket
type suspension insulator.**

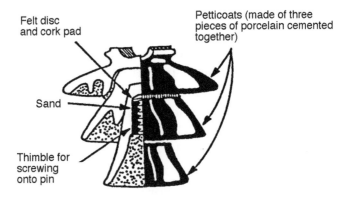

Figure A-1 b. Pin-type insulator.

Table A-1. Polymer insulation weight advantage

Product	Type	Voltage (kV)	Porcelain Weight (lbs)	Polymer Weight (lbs)	Percent Weight Reduction
Insulator	Distribution	15	9.5	2.4	74.7
Arrester	Distribution	15	6.0	3.8	36.7
Post Insulator	Transmission	69	82.5	27.2	67.0
Suspension	Transmission	138	119.0	8.0	93.2
Intermediate Arrester	Substation	69	124.0	28.0	77.4
Station Arrester	Substation	138	280.0	98.9	64.7

Polymer insulation is generally associated with a mechanically stronger insulation, such as high strength fiberglass. The fiberglass insulation serves as an internal structure around which the polymer insulation is attached, usually in the form (and function) of petticoats (sometimes also referred to as bands, water shedders; but for comparison purposes, however, here only the term petticoat will be used). The insulation value of the Polymer petticoats is equal to or greater than that of the fiberglass to which it is attached.

The internal fiberglass structure may take the form of a rod (or shaft), a tube, cylinder, or other shape. It has a high comparable compression strength as a solid and its tensile strength, equally high, is further improved by stranding and aligning around a fiber center. The polymer petticoats are installed around the fiberglass insulation and sealed to prevent moisture or contamination from entering between the petticoats and fiberglass; Figure A-3. The metal fittings at either end are crimped directly to the fiberglass, developing a high percentage of the inherent strength of the fiberglass. It should be noted that fiberglass with an elastometric (plastic) covering has been used for insulation purposes since the early 1920's.

The polymer petticoats serve the same function as the petticoats associated with porcelain insulators, that of providing a greater path for electric leakage between the energized conductors (terminals, buses, etc.) and the supporting structures. In inclement weather, this involves the shedding of rain water or sleet as readily as possible to maintain as much as possible the electric resistance between the energized element and the supporting structure, so that the leakage of electrical current between these two points be kept as low as possible to prevent flashover and possible damage or destruction of the insulator. Tables A-2a&b.

Figure A-2. A variety of typical Polymer insulator shapes. (*Courtesy Hubbell Power Systems*)

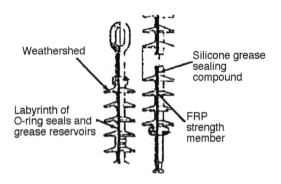

Figure A-3. Polymer insulator showing fiberglass rod insulation and sealing (Courtesy Hubbell Power Systems).

Table A-2a. Polymer improvement over porcelain (*Courtesy Hubbell Power Systems*)

	Watts Loss Reduction* (watts per insulator string)				
Voltage		*Relative Humidity*			
kV	*30%*	*50%*	*70%*	*90%*	*100%*
69	0.6-	0.8	0.9	1.0	4.0
138	1.0	2.4	4.5	7.2	8.0
230	1.0	2.5	5.7	14.0	29.0
345	2.5	4.2	8.5	15.0	30.0
500	2.8	7.8	11.5	33.0	56.0

*Power loss measurements under dynamic humidity conditions on I-strings.

Table A-2b. Polymer Distribution Arrester Leakage Distance Advantage (*Courtesy Hubbell Power Systems*)

Standard MCOV	Standard Porcelain Leakage Distance (in)	Special Porcelain Height (inches)	Special Porcelain Leakage Distance (in)	Standard Porcelain Height (inches)	Standard Polymer Leakage Distance (in)	Polymer Height (inches)
8.4	9.0	9.4	18.3	15.9	15.4	5.5
15.3	18.3	15.9	22.0	20.0	26.0	8.5
22.0	22.0	20.0	29.0	28.9	52.0	17.2

When the insulator becomes wet, and especially in a contaminated environment, leakage currents begin to flow on the surface; if the current becomes high enough, an external flashover takes place. The rate at which the insulation dries is critical. The relationship between the outer petticoat diameter and the core is known as the form factor. The leakage current generates heat (I^2R) on the surface of the insulator (eddy currents). In addition to the effects of the leakage current, the rate at which the petticoat insulation will dry depends on a number of factors. Starting with its contamination before becoming wet, the temperature and humidity of the atmosphere and wind velocity following the cessation of the inclement weather. In areas where extreme contamination may occur (such as some industrial areas or proximity to ocean salt spray), the polymer petticoats may be alternated in different sizes, Figure A-4, to obtain greater distance between the outer edges of the petticoats across which flashover might occur. When dry, the leakage current (approximately) ceases and the line voltage is supported across dry petticoats, preventing flashover of

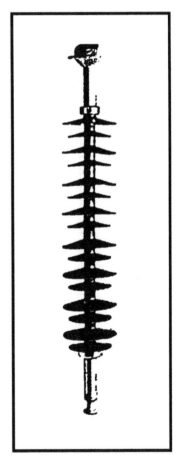

Figure A-4. Polymer insulator arrangement areas of high contamination where flashover between petticoat edges is possible (*Courtesy* Hubbe) Power Systems)

Table A-3. Comparison of Contamination Performance of Polymer versus Porcelain Housed Intermediate Class Arresters (Courtesy Hubbell Power Systems)

MCOV (kV)	Housing Material	Housing Leakage Distance (in)	Partial Wetting Test Max. Current (mA crest)	Max. Disc. Temps. (°C)	5 Hr. Slurry Test Maximum Current After Slurry Number			
					5	10	15	20
57	Polymer	81	<1	<38	35	42	44	44
66	Porcelain	54	68	>163	—	—	—	—
84	Polymer	109	<1	<38	50	52	60	60
98	Porcelain	122.4	18	<82	143	160	175	185

Tested using the 5-hour uniform slurry test procedure. This test consists of applying a uniform coating of standard 400 ohm-cm slurry to the test arrester. Within 30 seconds, MCOV is applied for 15 minutes, during which time the surface leakage currents cause the surface to dry. Slurry applications are repeated for a total of 20 test cycles. After the 20th test cycle, MCOV is applied to the arrester for 30 to 60 minutes to demonstrate thermal stability Surface leakage currents were measured at the end of the 5th, 10th, 15th and 20th test cycles.

the insulator. It is obviously impractical to design and manufacture comparable porcelain insulators as thin as polymers and having the same form factor. Table A-3.

In addition, in porcelain insulators, the active insulating segment is usually small and, when subjected to lightning or surge voltage stresses, may be punctured. In subsequent similar circumstances, it may breakdown completely, not only causing flashover between the energized element and the supporting structure, but may explode causing porcelain fragmentation in the process; the one-piece fiberglass insulator will not experience puncture.

The polymer suitable for high voltage application consists of these materials:

1. Ethylene Propylene Monomer (EPM)
2. Ethylene Propylene Diene Monomer (EPDM)
3. Silicone Rubber (SR)

Both EPM and EPDM, jointly referred to as EP, are known for their inherent resistance to tracking and corrosion, and for their physical properties, SR offers good contamination performance and resistance to Ultra Violet (UV) sun rays. The result of combining these is a product that achieves the water repellent feature (hydrophobic) of silicone and the electromechanical advantages of EP rubber.

Different polymer materials may be combined to produce a polymer with special properties; for example, a silicone EPDM is highly resistant to industrial type pollution and ocean salt.

The advantageous strength to weight ratio of polymer as compared to porcelain makes possible lighter structures and overall costs as well as permitting more compact designs, resulting in narrower right-of-way requirements and smaller station layouts. The reduction in handling, shipping, packaging, storage, preparation and assembly, all with less breakage, are obvious—these, in addition to the superior electromechanical performance.

Fiberglass insulation with its polymer petticoats is supplanting porcelain in bushings associated with transformers, voltage regulators, capacitors, switchgear, circuit breakers, bus supports, instrument transformers, lightning or surge arresters, and other applications. The metallic rod or conductor inside the bushing body may be inserted in a fiberglass tube and sealed to prevent moisture or contamination entering between the conductor and the fiberglass tube around which the polymer petticoats are attached. More often, the fiberglass insulation is molded around the conductor, and the polymer petticoats attached in a similar fashion as in insulators. Figure A-5.

Lightning or surge arrester elements are enclosed in an insulated casing. Under severe operating conditions, or as a result of multiple operations, the pressure generated within the casing may rise to the point where pressure relief ratings are exceeded. The arrester then may fail, with or without external flashover, Figure A-6, exploding and violently expelling fragments of the casing as well as the internal components, causing possible injury to personnel and damage to surrounding structures. The action represents a race between pressures building up within the casing and an arcing or flashover outside the casing. The 'length' of the casing of the arrester limits its ability to vent safely. The use of polymer insulation for the casing permits puncturing to occur, without the fragmentation that may accompany breakdown and failure of porcelain.

Summarizing, the advantages of polymers over porcelain include:

- Polymer insulation offers benefits in shedding rain water or sleet, particularly in contaminated environments.

Figure A-5. Typical porcelain bushings that may be replaced with polymers.

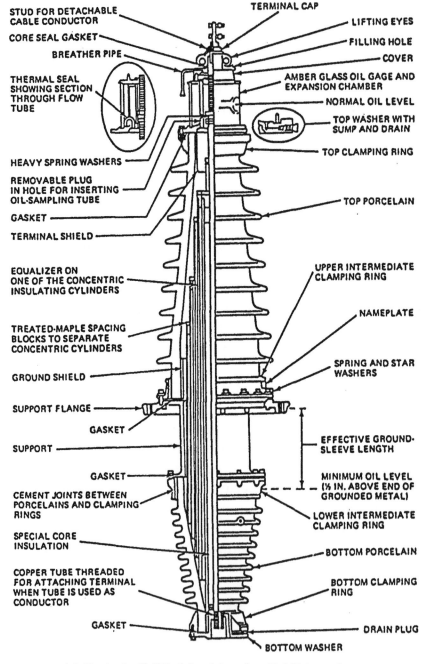

(a) Typical oil-filled bushing for 69 kV transformer.

Figure A-5(b). Sidewall-
mounted bushing.

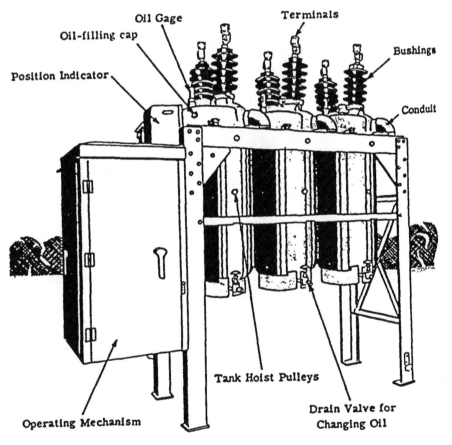

Oil circuit breaker.

Figure A-5(c).

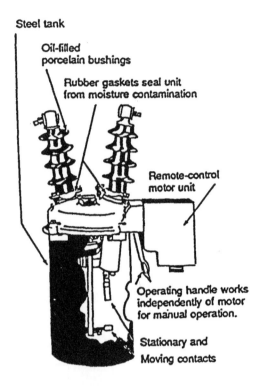

Figure A-5(d). Bushing applications that may be replaced with polymers.

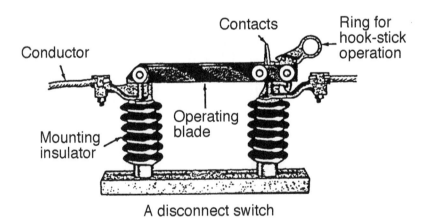

A disconnect switch

Figure A-5(e).

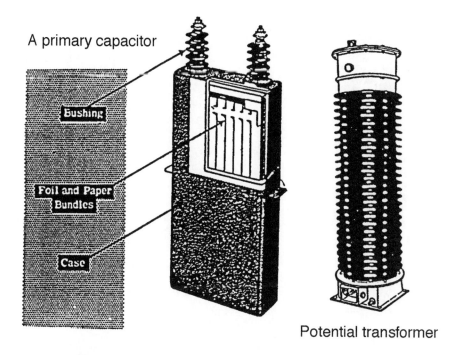

A primary capacitor

Bushing

Foil and Paper Bundles

Case

Potential transformer

Figure A-5f. More porcelain insulation that may be replaced with polymers.

• Polymer products weigh significantly less than their porcelain counterparts, particularly line insulators, resulting in cost savings in structures, construction and installation costs. Table A-4.

• Polymer insulators and surge arresters are resistant to damage resulting from installation and to damage from vandalism. The lack of flying fragments when a polymer insulator is shot deprives the vandal from his satisfaction with a spectacular event and should discourage insulators as convenient targets.

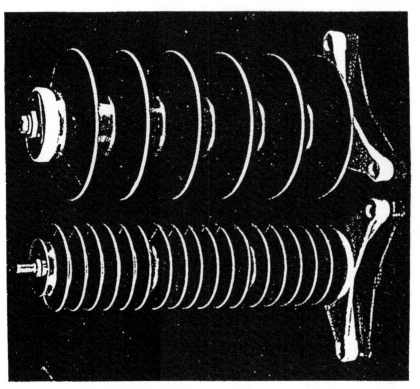

Polymer insulated arresters - station type

Photographs show the pressure relief capability of surge arresters.

Figure A-6. Polymer and porcelain cased arresters.

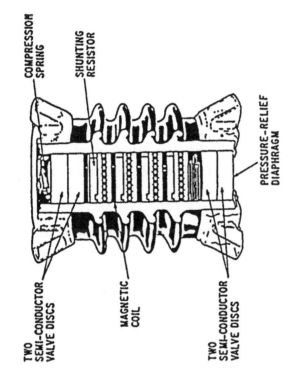

COMPRESSION SPRING

SHUNTING RESISTOR

PRESSURE-RELIEF DIAPHRAGM

TWO SEMI-CONDUCTOR VALVE DISCS

MAGNETIC COIL

TWO SEMI-CONDUCTOR VALVE DISCS

Cross-section of porcelain cased arrester

Polymer insulated intermediate type arrester

Figure A-6 (cont'd). Polymer and porcelain cased arresters.

- Polymer arresters allow for multiple operations (such as may result from station circuit reclosings), without violently failing. Figure A-6.

- Polymer insulators permit increased conductor (and static wire) line tensions, resulting in lower construction designs by permitting longer spans, fewer towers or lower tower heights.

- Polymer one-piece insulators, lacking the flexibility of porcelain strings and the firm attachment of the conductor it support are said to produce a tendency to dampen galloping lines.

Although polymer insulation has become increasingly utilized over the past several decades, there are literally millions of porcelain insulated installations in this country alone; economics does not permit their wholesale replacement. Advantage is taken of maintenance and reconstruction of such facilities to make the change to polymers.

Much of the data and illustrations are courtesy of Hubbell Power systems, and is herewith duly acknowledged with thanks.

Table A-4. Example—10 Miles, 345 kV, 250 Strings of Insulators

Porcelain -	4500 bells, 52-3, 13.5 lbs. ea., total 60,760 lbs.
	750 crates at 3.1 cu. ft. = 2,325 cu. ft.
	Insulator cost = $51.760 ($11.60/bell)
Polymers -	250 units, 14.4 lbs. ea., total 3 600 lbs.
	5 crates at 75 cu. ft. = 375 cu. ft.
	Insulator cost = $51.750

Savings

1. Storage space at receiving point (3 mos.)
   porcelain - 580 sq. ft.; polymer - 100 sq. ft . ... 480 sq. ft. $ 60.00
2. Off-load, re-load at receiving point; porcelain -
   10 man-hrs.; polymer - 2 man-hrs. .... 8 man-hrs. $120.00
3. Breakage - off-loading, storage; re-load-
   porcelain - 1 percent; polymer - 0 .... 1 percent $517.50
4. Truck - receiving point to tower sites (5 miles);
   porcelain- 1.00/cwt.; polymer .50/cwt $589.50 $589.50
5. Off load at tower site porcelain -
   5 man-hrs.; polymer - 1 man-hr . ... 4 man-hrs. $60.00
6. Unpack at tower site; porcelain -
   50 crates/hour, 25 man-hrs.; polymer -
   50 insulators/hour, 5 man-hrs . ... 20 man-hrs. $300.00
7. Breakage - off-loading through string assembly
   & cleaning porcelain - 1 percent: polymer - 0 .... 1 percent $517.50
8. Assemble strings, attach blocks porcelain -
   40 man-hrs.; polymer - 8 man-hrs . ... 32 man-hrs. $480.00
9. Clean insulators; porcelain-10 min./string;
   polymer - 3 min./string .... 29 man-hrs. $435.00
10. Lift string into place (2 men); porcelain -
    5 min./string; polymer - 2 min./string .... 25 man-hrs. $375.00
11. Install & connect to tower (2 men); porcelain -
    5 min./string polymer - 2 min./string .... 25 man-hrs. $375.00
12. Breakage - lifting & installation;
    porcelain - 0.5 percent; polymer - 0 .... 0.5 percent $258.75
13. Cleanup packaging materials at jobsite;
    porcelain - 6 man-hrs.; polymer - 1.5 man-hrs ...... 4.5 man-hrs. $67.50

(*Courtesy* Hubbell Power Systems)

Appendix B

The Grid
Coordinate System:
Tying Maps to Computers*

INTRODUCTION

The grid coordinate system is the key that ties together
two important tools, maps and computers. Maps are a necessity for the better operation of many enterprises, especially
of utility systems. Their effectiveness can be increased many-
fold by adding to their information data contained in other
files. Much of the latter data are now organized and stored
in computer-oriented files-on punched cards and on magnetic tapes, drums, disks, and cells. Generally, these data can
be retrieved almost instantly by CRTs (cathode ray tubes) or
printouts. The link that makes the correlation of data contained on the maps and in the files practical is the grid coordinate system.

Essentially, the grid coordinate system divides any particular area served into any number of small areas in a grid
pattern. By superimposing on a map a system of grid lines,
and assigning numbers to each of the vertical and horizontal
spacings, it is possible to define any of the small areas by
two simple numbers. These numbers are not selected at ran-

*Reprinted (with modifications) from *Consulting Engineer,*® January 1975, vol. 44, no. 1, pp.
5 1-55.© by Technical Publishing, a company of the Dun & Bradstreet Corporation, 1975.
All rights reserved.

dom, but have some meaning. Like any graph, these two coordinates represent measurements from a reference point; in this respect they are similar to navigation's latitude and longitude measurements.

Further subdivision of the basic grid areas into a series of smaller grids is possible, each having a decimal relation with the previous one (i.e., by dividing each horizontal and vertical space into tenths, each resultant area will be one-hundredth of the area considered). By using more detailed maps of smaller scale, it is possible to define smaller and smaller areas simply by carrying out the coordinate numbers to further decimals. For practical purposes, each of these grid areas should measure perhaps not more than 25 ft by 25 ft (preferably less, say 10 ft by 10 ft) and should be identified by a numeral of some 6 to 12 digits.

For example, by dividing by 10, an area of 1,000,000 ft by 1,000,000 ft (equivalent to some 190 miles square) can be divided into 10 smaller areas of 100,000 ft by 100,000 ft each, identified by two digits, one horizontal and one vertical. This smaller area can again be subdivided into 10 smaller areas of 10,000 ft by 10,000 ft each, identified by two more digits, or a total of four with reference to the basic 1,000,000-ft square area. Breaking down further into 1000- by 1000-ft squares and repeating the process allows these new grids to be identified by two more digits, or a total of six. Again dividing by 10 into units of 100 ft by 100 ft, and adding two more digits, produces a total of eight digits to identify this grid size. One more division produces grids of 10 ft by 10 ft and two more digits in the identifying number-for a total of 10 digits, not an excessive number to be handled for the grid size under consideration; see Figure B-1.

This process may be carried further where applications requiring smaller areas are desirable; however, each further breakdown not only reduces the accuracy of the measure-

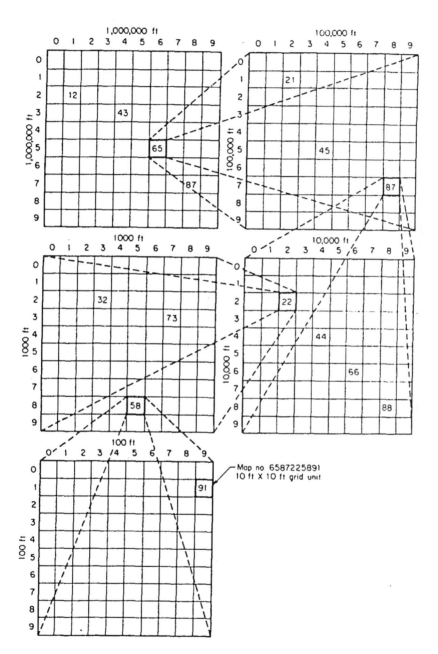

Figure B-1. Development of grid coordinate system.

ments, but also adds to the number of digits, which soon becomes unmanageable. Experience indicates that a "comfortable" system should contain 10 digits or fewer for normal usages.

While the decimal relation has been mentioned, other relations can be used, such as sixths, eighths, etc., or combinations, such as eighths and tenths, and others.

Standard References

To give these numerals some actual physical or geographical significance, they may be tied in with existing local maps. U.S. Geological Survey maps, coast and geodetic survey maps, state plane coordinate systems, standard metropolitan statistical areas, or latitude and longitude bearings. They may also be tied in with maps independent of all of these.

While reference to state and federal government systems lends some geographical significance, it produces identifying grid numbers with several additional digits. It is not necessary for any grid coordinate system to have this reference to a government system, but if it is desired, it is a relatively simple procedure to develop a computerized look-up program that can translate such coordinates.

Basic grid coordinate maps may be developed from the conversion of existing maps to a usable scale, if such maps are reasonably accurate and complete, both as to their geography and content. They may also be developed from exact land surveys, from aerial surveys, or from combinations of all of these.

Excellent maps are also available for most of the country. U.S. Geological Survey maps show latitude and longitude lines every few miles; they also show numerous triangulation stations with the latitude and longitude for each station determined to an extreme degree of accuracy.

Further, detail maps are available for practically every city and township, showing streets, houses, and lots. Despite the fine degree of accuracy of these maps, minor inaccuracies and discrepancies are bound to occur.

Earth's Curvature

Errors occur in mapping the earth's curved surface on a flat map; see Figure B-2. For example, in the case of the approximate 190-mi square mentioned previously, in the continental United States, the error introduced by this curvature, measuring from the center (95 mi in the longitudinal, or north-south direction) would probably not exceed 2 percent, a tolerable error. These errors need not be of great import, except in establishing match lines between maps. No gaps or

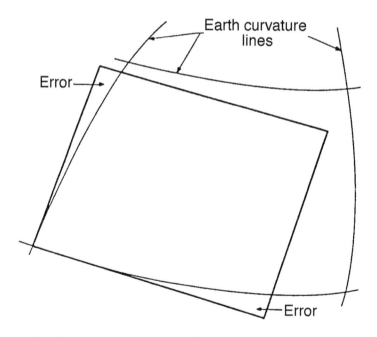

Figure B-2. Error introduced in grid coordinate system by earth curvature

overlaps should appear between adjacent maps, or between property or lot lines within a map. Tolerances of a few percent ordinarily are acceptable.

GRID COORDINATE MAPS

A grid coordinate map system should meet the following requirements:

1. It should include a simple and easily understood system of numerals for locating the data under consideration (numerals only; the x and y coordinates).

2. The grid areas should be small enough to be consistent with the purposes for which they are to be used (25 ft or less).

3. The number of digits in the grid number should be held to a practical number so as not to become cumbersome and unwieldy (normally not more than about 10).

4. It should be designed to allow for expansion so that it will not have to be radically revised if unforeseeable expansion should occur.

5. It must provide reasonable accuracy (tolerances of a few percent), and it may or may not be tied in with some government or other established coordinate system.

6. Map sizes should be manageable (say, 24- or 36-in. square).

7. Maps of different scales should be included in the sys-

tem to accommodate different kinds of data (for circuit data, say 1000 or 500 ft; for details of facilities, say, 100 ft for overhead and 50 or 25 ft for underground).

8.   A key map must show the entire grid area.

9.   Optional is a grid atlas showing street locations with grid overlay.

In attempting to design a grid coordinate system for a very large area, it may be difficult to meet these requirements. In such instances it may be desirable and practical to divide the entire area into two or more convenient districts, establishing a separate grid coordinate system in each district and tying the separate systems together with match lines at the borders. A prefix letter or number may be used to identify districts, though this may not prove necessary in actual operation.

The size of individual maps should be large enough to encompass an area suitable for the purpose but small enough not to be unwieldy; sizes 24- or 26-in. square have proven practical. Maps of different scales are used for different purposes; for example, a 50- or 100-ft scale is used for dense or crowded areas; 300- or 500-ft for less dense or rural areas; and 500- or 1000-ft or even larger for district or overall area viewing. The series of scales used should be such that the larger-scale maps fit into those of smaller scale completely and evenly. Match lines of each sheet should fall on corresponding match lines of adjacent sheets.

The grid pattern applicable to each of the several scale maps may be printed directly on each map as light background lines, perhaps even in a different color, or printed on the back of the maps when they are reproduced. Alternately, a grid overlay can be applied to each map to be used when it is necessary to determine a grid coordinate for an item on the map. The actual grid number need not be printed on every item on every map unless desired. Such numbers assigned to key locations on each map normally suffice; the others may be determined from the grid background or overlay. To maintain their permanence and to minimize distortion from expansion and contraction because of changing humidity and temperature, the maps should be printed on a material such as Mylar, a translucent polyesterbase plastic film; this is especially true of the base maps from which others of different scales and purposes are derived.

As much as practical, the data on the maps should be uncluttered and as legible as possible. It may be desirable in some instances to provide two or more maps (of the same scale) for several purposes; marks for coordinating these several maps, should it be necessary, may be included on each of the maps.

Maps may be further uncluttered by deliberately removing as much of the information on them as appears desirable and practical and consigning such information to files readily accessible by computer. In many instances, this information already is included or duplicated in such files, but may need to be labeled with the appropriate grid coordinate number. The use of CRTs and printouts makes this information available at will.

Application of Grid Coordinate Numbers

The grid coordinate number may be applied to each item of information contained in the computer-operated files by location. This may be done in several ways: manually, by machine, or a combination of the two.

The manual method is to superimpose or overlay the grid pattern on existing maps and manually assign numbers to each item to be processed. As mentioned previously, the grid pattern may be transferred to the master or original maps, and reproduced (or microfilmed) on the maps for the user; here numbers can be assigned directly from the map.

The machine method of grid number assignment employs an electronic scanning device called a *digitizer*. This machine includes a drafting table for map display and a cursor or pointer. The postal address and other fixed data are inserted on a punch card. For a particular map, the grid numbers of the map are set on the digitizer console. The digitizer assigns the x and y coordinates when the cursor is placed on a selected point and activated. These data are fed to a keypunch, which produces a punched card. In this method of producing the grid coordinate numbers, the digitizer enables additional refinement to be achieved, producing additional decimal numbers for the x and y coordinates. Hence, the ultimate grid area that can be measured can be one-tenth or one-hundredth, etc., of the basic unit area (1 by 1 ft, or 0.1 ft by 0.1 ft, for a 10- by 10-ft base area).

When numbers are assigned manually from maps, this degree of accuracy is not possible, nor is it necessary if the principal purpose of the grid coordinate system is to identify an item rather than a precise point. While the actual accuracy of such additional digits can be questioned, they provide a method of further subdividing a map for closer location of an item, but more important, they make possible a system of automatic mapping using the computer.

The grid coordinate number corresponding to the location of an item in question is added to its record and now becomes its computer *address*. In assigning these numbers to existing files, the digitizer generally can properly identify the location from a suitable map. As a practical matter, however, there will be some locations or descriptions that cannot be identified using the digitizer, and these may require manual processing and actual checking in the field; fortunately these usually constitute only a small percentage of the total records.

In the maintenance of such files, the grid coordinate numbers associated with changes in, or with the introduction of new items into, the records can be assigned manually by the originator of the record.

COORDINATE DATA HANDLING

As implied earlier, the grid coordinate system provides an easy and simple but, more important, a very rapid means of obtaining data from files through the use of the computer. In some respects, it assigns addresses to data in the same way as the ZIP code system in use by the postal service. The manner in which the grid number may be used is illustrated in the following examples; for convenience they refer to electric utility systems, although obviously they apply equally well to other endeavors employing maps and records.

Data contained on maps and records generally apply to the consumers served and the facilities installed to serve them. While maps depict (by area) the geographic and functional (electrical) interrelationship between these several components, the records supply a continuing history (by location) of each component item (consumers and facilities).

In the case of consumers, such data may include, in addition to the grid coordinate number, the name and post office address. Also a history of electric consumption (and demand where applicable), billing, and other pertinent data over a continuing period, usually 18 or 24 months. There may also be data on the consumer's major appliances; also the data and work order number of original connection and subsequent changes. The grid coordinate number of the transformer from which the consumer is supplied is included, as well as that for the pole or underground facility from which the service to the consumer is taken. Sometimes interruption data may be included. Other data may include telephone number, tax district, access details, hazards (including animals), dates of connection or reconnection, insurance claims, easements, meter data, meter reading route, test data, credit rating, and other pertinent information. Only a small portion of these data are shown on maps, usually in the form of symbols or code letters and numerals. In the case of facilities, such data may include, in addition to the grid coordinate number, location information, size and kind of facility (e.g., pole, wire, transformer, etc.), date installed or changed, repairs or replacements made (including reason therefor, usually coded), original cost, work order numbers, crew or personnel doing work, construction standard reference, accident reports, insurance claims, operating record, test data, tax district, and other pertinent information. Similarly, only a small portion of these data are shown on maps, usually in the form of symbols or code letters and numerals.

Data from other sources also may be filed by grid number for correlation with consumer and facility information for a variety of purposes. Such data may include government census data; police records of crime, accidents, and vandalism; fire and health records; pollution measurements; public planning; construction and rehabilitation plans; zoning restrictions; rights-of-way and easement locations; legal data; plat and survey data; tax district; and much other information that may affect or be useful in carrying out utility operations.

Obviously, all these data, whether pertaining to the consumer or to the utility's facilities, are not necessarily contained on one map or in one record only; indeed, there may be several maps and records involved, each containing certain amounts of specialized or functionally related data. All, however, may be correlated through the grid coordinate system.

Data Retrieval

Data contained in the files may be retrieved by means of the computer and may be presented visually by means of CRTs for one-time instant use, or by printouts and automatic plotting for repeated use over an indefinite time period. Data presented may be the exact original data as contained in one or more files, or extracted data obtained as a result of correlating data residing in one or more files, or a combination of both; such extracted data may or may not be retained in separate files for future use.

These data may be retrieved for an individual consumer or an individual item of plant facilities, or may be other data for a particular area, small or large. The various specific purposes determine what data are to be retrieved and how they are to be presented. They also determine the programs and equipment required. Data thus retrieved then are used with data contained on the map to help in forming the decisions required. The decisions may include new data that can be reentered in the files as updating material, that can be plotted or printed for exhibit purposes, or that can be reentered on maps for updating or expanding the material thereon; all of these may be done by means of the computer.

The grid coordinate number is applied to utility facilities for ease of location and positive identification in the field. In the case of electric utilities, these may include services, meters, poles, towers, manholes, pull boxes, transformers, transformer enclosures, switches, disconnects, fuses, lightning arresters, capacitors, regulators, boosters, streetlights, air pollution analyzers, and other equipment and apparatus; also the location of laterals on transmission and distribution circuits.

OTHER APPLICATIONS

Similarly, for gas utilities, the applications of grid coordinate numbers may include mains, services, meters, regulators, valves, sumps, test pits, and other equipment; also the location of boosters, laterals, and nodes on the gas systems. For water systems, they may include mains, services, meters, valves, dams, weirs, pumps, irrigation channels, and other facilities. For telephone and telegraph communication systems, including CATV circuits, they may include mains, services, terminals, repeaters, microwave reflectors, and other items including poles, manholes, and special items.

Grid coordinate numbers also may find application in many other lines of endeavor: highway systems, railway systems, oil fields, social surveys (police, health, income, population distribution, etc.), market surveys (banks and industries), municipal planning and land use studies, non-classical archeology, geophysical studies, and others where such means of location identification may prove practical.

The use of grid coordinates facilitates positive identification in the field; the numbers are posted systematically on facilities, such as streetlight or traffic standards, poles, and structures, and at corners or other prominent locations.

An atlas, consisting of a grid overlay on a geographical map, aids the field forces in locating consumers and plant facilities and provides a common basis for communication between office and field operating personnel.

The grid pattern permits the classical manipulation of data by individual grid sections or areas comprising several grid sections. In addition to the sample presentation of such data by means of CRT displays and typed printouts, data may be presented in the form of plotting in various graphical forms, in patterns indicating the distribution of data, the density of particular data, the accumulation of data within fixed boundaries, the determination of area boundaries for predetermined data content (the analysis of data within a given polygon), the calculation of lengths and distances between grid locations, and the mapping of facilities in acceptable detail-and all of these operations may be performed automatically by means of the computer.

Further, summaries and analyses employing the grid coordinate system may be more readily made and are susceptible to combination and consolidation, resulting in perhaps fewer and more comprehensive reports (eliminating the duplication of much needless data and the presentation of more complete and meaningful conclusions in one place).

In all of the foregoing discussion, the point must be made that all of the handling of data using the grid system may also be accomplished without the use of the grid system. It is apparent, however, that this latter method will in the vast majority of cases employ more effort in terms of work hours and will be more time-consuming, so as to render many applications impractical, even though their desirability may be great; in short, the grid coordinate system enhances the economics of data handling.

ECONOMICS

It is not to be denied that the introduction of the grid coordinate system will impose additional cost to the maps and records function. It is also evident that these costs will be offset by the decreased personnel requirements in the processing of data derived from the maps and records, especially when the computer may be made to take up a large part of this burden. Moreover, more refinement and a wider scope in processing of data are attainable.

The cost of implementing a grid coordinate system can

be evaluated fairly accurately. Many factors will influence the final determination; these include the area of the system involved, the number of consumers and facilities, the condition of the basic and auxiliary maps and records, the number and scope of the applications desired, the extent of automation, and many other factors. A very approximate estimate may average perhaps about one day's revenue per consumer. Practical considerations associated with implementation may well dictate a period of several years, perhaps 5 years or even more, over which the expenditure will have to be made to accomplish the desired goals.

The offsetting savings from the introduction of a grid coordinate system, including those derived from the additional worth of the wider utilization, are difficult to pinpoint. It should be observed that while it is probable that a single application will not justify the adoption of the grid coordinate system, except in some unusual or special set of circumstances, it is also probable that the multiplicity of practical applications indicated will justify the relatively modest expenditure necessary for the conversion of present maps and records to the grid coordinate system.

The personnel requirements necessary to implement a grid coordinate system over a reasonable (short-term) period of time must be viewed together with the overall probable lessened longer-term in-house requirements. Since such a conversion is a one-time operation, it recommends itself admirably to the classical use of contractors having the special skills and experience. Further, such outside services are not apt to be diverted by crisis incidents prevalent in many enterprises.

One final observation. With the national consensus apparently pointing to an ultimate metric system for the United States to conform with world standards, the adoption of a grid coordinate system provides an excellent opportunity for its introduction with a minimum of conversion effort.

With the advent of the computer, it was inevitable that the grid coordinate system should be developed to provide a simple means of addressing the computer. The grid number provides the link between the map and the vast amount of data managed by the computer. This happy marriage of two powerful tools results not only in better operations but in improved economy as well. It is a must in the modernization of operations in many enterprises and especially in utility systems.

Appendix C

US and Metric Relationships

U.S. To Metric		Metric To U.S.	
Length			
1 inch	= 25.4 mm	1 millimetre	= 0.03937 inch
1 inch	= 2.54 cm	1 centimetre	= 0.3937 inch
1 inch	= 0.0254 m	1 metre	= 39.37 inch
1 foot	= 0.3048 m	1 metre	= 3.2808 feet
1 yard	= 0.9144 m	1 metre	= 1.094 yard
1 mile	= 1.609 km	1 kilometre	= 0.6214 mile
Surface			
1 inch^2	= 645.2 mm^2	1 millimetre^2	= 0.00155 inch^2
1 inch^2	= 6.452 cm^2	1 centimetre^2	= 0.155 inch^2
1 foot^2	= 0.0929 m^2	1 metre^2	= 10.764 foot^2
1 yard^2	= 0.8361 m^2	1 metre^2	= 1.196 yard^2
1 acre	= 0.4047 hectare	1 hectare	= 2.471 acres
1 mile^2	= 258.99 hectare	1 hectare	= 0.00386 mi^2
1 mile^2	= 2.59 km^2	1 kilometre^2	= 0.3861 mile^2
Volume			
1 inch^3	= 16.39 cm^3	1 centimetre^3	= 0.061 inch^3
1 foot^3	= 0.0283 m^3	1 metre^3	= 35.314 foot^3
1 yard^3	= 0.7645 m^3	1 metre^3	= 1.308 yard^3
1 foot^3	= 28.32 litres	1 litre	= 0.0353 foot^3
1 inch^3	= 0.0164 litre	1 litre	= 61.023 inch^3
1 quart	= 0.9463 litre	1 litre	= 1.0567 quarts
1 gallon	= 3.7854 litres	1 litre	= 0.2642 gallons
1 gallon	= 0.0038 m^3	1 metre^3	= 264.17 gallons
Weight			
1 ounce	= 28.35 grams	1 gram	= 0.0353 ounce
1 pound	= 0.4536 kg	1 kilogram	= 2.2046 lb*
1 net ton	= 0.9072 T (metric)	1 Ton (metric)	= 1.1023 net tons**
*Avoirdupois			
**1 ton = 2000 lb			
Compound units			
1 lb/ft	= 1.4882 kg/m	1 kilogram/metre	= 0.6720 lb/ft
1 lb/in^2	= 0.0703 kg/m^2	1 kg/cm^2	=14.223 lb/in^2

Appendix D

Transmission Planning and The Need for New Capacity

(National Transmission Grid Study)

Eric Hirst
Consulting in Electric-Industry Restructuring
Oak Ridge, Tennessee
Brendan Kirby
Oak Ridge National Laboratory
Oak Ridge, Tennessee

INTRODUCTION

The U.S. electricity industry is in the midst of a transition from a structure dominated by vertically integrated utilities regulated primarily at the state level to one dominated by competitive markets. In part, because of the complexities of this transition, planning and construction of new transmission facilities are lagging behind the need for such grid expansion.

Between 1979 and 1989, transmission capacity increased slightly faster than did summer peak demand (Hirst and Kirby 2001). However, during the subsequent decade, utilities added transmission capacity at a much lower rate than loads grew. The trends established during this second decade are expected to persist through the next decade. According to one analysis, maintaining transmission adequacy at its current level might require an investment of about $56 billion during the present decade, roughly half that needed for new generation during the same period (Hirst and Kirby 2001).

Expanding transmission capacity requires good planning (as well as appropriate market rules and regulatory oversight). The Federal Energy Regulatory Commission (FERC 1999) emphasized the importance of transmission planning in the creation of competitive wholesale markets. FERC wrote that each regional transmission organization (RTO) "must be responsible for planning, and for directing or arranging, necessary transmission expansions, additions, and upgrades that will enable it to provide efficient, reliable, and nondiscriminatory transmission service and coordinate such efforts with appropriate state authorities." FERC included transmission planning as one of the eight minimum functions of an RTO:

[T]he RTO must have ultimate responsibility for both transmission planning and expansion within its region that will enable it to provide efficient, reliable and non-discriminatory service.... The rationale for this requirement is that a single entity must coordinate these actions to ensure a least cost outcome that maintains or improves existing reliability levels. In the absence of a single entity performing these functions, there is a danger that separate transmission investments will work at cross-purposes and possibly even hurt reliability.

This shift from planning conducted by individual utilities for their system to meet the needs of their customers, to planning conducted by RTOs to meet the needs of regional electricity markets, raises important issues (Table 1). These issues include the criteria for planning (reliability, economics, etc.); environmental considerations (effects of transmission expansion on the location and types of emissions from power plants, accommodation of remotely located renewable resources, as well as the direct siting and environmental effects of transmission); economic development (providing greater access to cheaper power may encourage local and regional economic growth); the role of congestion costs in deciding which projects to build; the consideration of generation, load, and transmission-pricing alternatives to new transmission projects; the economic and land-use benefits of building larger facilities ahead of immediate need; the role of new solid-state technologies that permit operation of transmission systems closer to their thermal limits; the role of merchant transmission projects; and the growing difficulty in obtaining data on new generation and load growth caused by the separation of generation and retail service from transmission. Finally, collaborative transmission-planning processes, which include various stakeholders early in the

Table 1. Key transmission-planning issues

Topic	Issues
Reliability vs commerce	To what extent should RTOs plan solely to meet reliability requirements, leaving decisions on grid expansion for commercial purposes (e.g., to reduce congestion costs) in the hands of market participants?
Congestion costs	Are congestion costs (e.g., short-term nodal or zonal congestion prices and long-term firm transmission rights) a suitable basis for deciding on transmission investments?
Alternatives to transmission	What role should RTOs play in assessing and motivating suitably located generation and load alternatives to new transmission? Should RTOs provide information only or should they also help pay for such alternatives?
Economies of scale	Should RTOs or private investors overbuild transmission facilities in anticipation of future need to reduce the dollar and land costs per GW-mile of new transmission facilities? How should these economies be balanced against the possibly greater financial risks of larger transmission facilities?
Advanced technologies	What are the prospects for widespread use of new technologies (e.g., superconductivity, solid-state electronics, and faster systems to collect and analyze data) to improve system control, thereby permitting operation of existing grids closer to their limits?

(Continued)

Table 1. Key transmission-planning issues (*Continued*)

Planning data	Who will provide the data needed for transmission planning, particularly on the locations, timing, and types of new and retiring generating units and the loads and load shapes of retail customers?
Economic effects	How should transmission's impact on regional power prices and the resulting impact on the regional economy be factored into transmission planning?
Environmental and other societal effects	How should the effects of transmission availability on the generation mix and the resulting shift in emissions be included in transmission planning? How should remotely located generators (e.g., coal and wind) be accommodated in transmission planning? Should transmission be built to increase fuel diversity for generation and to discipline generator market power? How should potential siting problems be incorporated into the planning process?
Centralized vs decentralized transmission planning and expansion	To what extent can private investors, rather than RTO planners, decide on and pay for new transmission facilities? Can they, in spite of networkexternality effects, capture enough of the benefits of such transmission projects to justify their investment? How can new technologies advance private investment?

process (e.g., as problems are being identified rather than when solutions have already been selected), should be considered as RTOs plan for future regional electricity needs.

Part of the complexity associated with transmission planning stems from transmission's central position in electric-system operations and wholesale power markets. Because of its centrality, transmission serves many commercial and reliability purposes. American Transmission Company (2001) identified several objectives for transmission planning and expansion: improve transfer (import and export) capability from different directions, accommodate load growth without delay, accommodate generation development without delay, provide flexibility to transmission customers to modify their transactions as market conditions change, reduce service denials and interruptions due to transmission constraints (equivalent to reducing congestion costs), cut losses, and improve reliability. Southern Company Services (1995) mentions many of the same objectives and also includes provision of sufficient margin to permit transmission elements to be taken out of service temporarily for maintenance.

The American Transmission Company (2001) plan notes some of the many issues it will have to consider as it plans for transmission expansion, including public involvement in the planning process, minimizing environmental and land-use impacts, timely licensing and construction of good projects, and balancing the robustness of the transmission system with the need to keep transmission rates reasonable.

The foregoing comments on the purposes and complexities of transmission planning emphasize the fact that such planning is only one element of a broader process that ultimately leads to the construction of needed bulk-power facilities (Figure 1). To assess various transmission and nontransmission (generation, load, and pricing) alternatives,

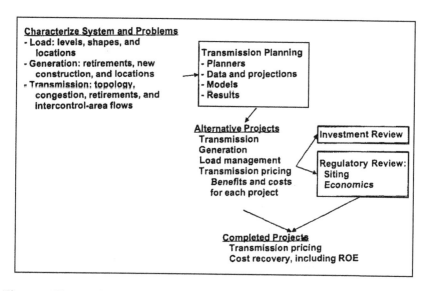

Figure 1.Transmission-planning models, and their inputs and outputs.

transmission models require large amounts of data and projections related to loads, generation, and transmission. Transmission planners use detailed electrical-engineering
computer models to assess these alternatives (Figure 2).
Model results, combined with information on project costs,
environmental effects, siting, and regulatory requirements,
lead to financial and regulatory assessments of different
projects. Ideally, these plans lead to the construction of
needed projects, cost recovery (including a return on investment) for transmission owners, and transmission rates that
appropriately charge users for the services they receive.

Figure 2 expands on the transmission-planning portion
of Figure 1. This second figure shows how load-flow, dynamic, and short-circuit models are used to determine
whether the bulk-power system can meet all the applicable
operating and planning reliability standards. The arrow to
the right of the box labeled Planning Models indicates that
these models are run over and over to test the ability of the

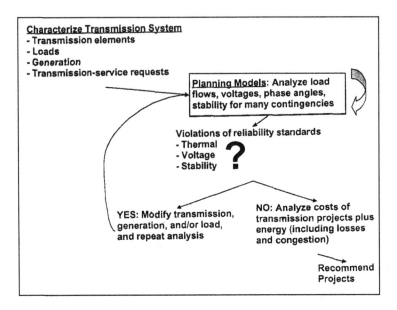

Figure 2. The relationship between transmission planning and its inputs (data and projections) and results.

bulk-power system to operate within specified ranges for all first- and some multiple-contingency conditions. The fundamental characteristic that makes transmission planning and investment so difficult is lack of control of the grid and the inability to control the flow through individual transmission elements (e.g., lines and transformers). (Devices such as phase shifters and direct current (DC) links allow control, but are much more expensive than traditional transmission facilities.) Each transmission element is part of a network that is a common resource available to all. Because electricity flows according to the laws of physics and not in response to human controls, what happens in one part of the grid can affect users throughout the grid. Because of these large externalities, transmission must be centrally managed and regulated. Other characteristics that complicate transmission planning include:

- Large Geographic Scope—Conditions on one part of an alternating current (AC) network affect flows throughout the network. Consequently, transfers between any two points on the network can be restricted by constraints elsewhere in the network. Similarly, upgrades to any part of the network affect transfer capabilities throughout the network.

- Diversity of Interests—Each transmission enhancement affects many market participants. Generators will either expand their market opportunities (if they are low-cost producers) or reduce their market opportunities. Loads have similar, but opposite, interests.

- Transmission vs Generation—The split and differences between competitive generation and regulated transmission affect transmission planning. The competitive generation business encourages faster planning, shorter deployment times, and less sharing of commercially sensitive information. The regulated transmission business environment produces slower planning and longer deployment times (to accommodate an inclusive public process) and the wide sharing of information. In addition, transmission and generation are both complements and substitutes. As a consequence, poor transmission planning and inefficient transmission expansion could undercut competitive wholesale markets and increase electricity costs.

- Long Life—Transmission is a long-lived (30 to 50 years), immobile investment with very low operating costs. The need for new transmission shows up in real-time congestion prices. It is difficult to accurately forecast the need for a specific transmission investment for several decades. The generation and demand-side alternatives are often shorter lived and have higher operating costs that can be eliminated if the investment is no longer needed.

- Regulatory Decision Process—Because the regulator (and the regulated entity) are spending ratepayer dollars, public processes are used to produce good decisions. All opinions and options are welcome and considered, which can lead to a time-consuming and costly process.

- Regulatory Uncertainly—Investors are unlikely to spend their money until it is clear that they will recoup their investment and earn a reasonable return on that investment.

- Environmental Impacts—Some people oppose new transmission lines (and, to a lesser extent, substations) on aesthetic grounds or because they might lower property values. Others are concerned about the health effects of electromagnetic fields. Although little scientific evidence supports this concern about transmission lines, public perceptions and fears may lead to opposition to construction of new transmission lines (National Institute of Environmental Health Sciences, 1999).

The remainder of this issue paper is organized as follows: The next section summarizes planning processes as practiced by vertically integrated utilities and today's independent system operators (ISOs). This section also summarizes the planning processes proposed by RTOs. Subsequent sections outline the characteristics of an ideal transmission plan and planning process; a benchmark against which current and future plans might be assessed; and several key planning issues and the complications that arise because of the increasing competitiveness and transitional state of the U.S. electricity industry. A later section recommends certain actions for DOE, FERC, and others on improved planning processes; while the final section summarizes the key findings from this issue paper.

TRANSMISSION PLANNING PRACTICES

Traditional Utilities

Historically, transmission planning was much simpler than it is today and than it is likely to be in the future. Until the mid-1990s, the U.S. electricity industry featured vertically integrated utilities. As a consequence, transmission planning was closely coupled to generation planning. Utilities, because they owned generation and transmission, could optimize investments across both kinds of assets. With respect to operations, utilities routinely scheduled generation day-ahead and redispatched generating units in real time to prevent congestion from occurring. The costs of such scheduling and redispatch were spread across all customers and reflected in retail rates.[1]

In addition, utilities had good data and forecasting tools to estimate current and future loads and generating capacity. Because each utility was the sole provider of retail electricity services, it had considerable information on current and likely future load levels and shapes. Because each utility was

---

[1]Although transmission planning focused primarily on generation and loads within a single control area, the tight power pools and regional reliability councils reviewed utility plans to ensure that projects proposed in one service area would not adversely affect other utility service areas.

the primary investor in new generation, it had considerable information on the timing, types, and locations of new generation and corresponding information on the retirement of existing units.

Finally, the amount of wholesale electricity commerce was much less than it is today and it was much simpler. It was simpler in the sense that most transactions involved neighboring utilities, either to take advantage of short-term economies of operation or for long-term purchases of firm power.

Current Planning Environment

In today's electricity industry, generation and transmission are increasingly separated, either through functional unbundling of these activities or through corporate separation. This deintegration, combined with the competitive nature of electricity generation, makes it much harder for transmission planners to coordinate their activities with those of generation owners. In particular, the owners of generation are reluctant to reveal their plans for new construction and retirement of existing units any sooner than they have to.

In some regions, today's system operators are independent of load-serving entities. Therefore, the system operators have little information on the details of retail loads, such as the types of end-use equipment in place and trends and patterns in electricity use. It is now the load-serving entities that have such information, and for competitive reasons, they may be reluctant to share such information and projections with the system operator.

This deintegration of generation and transmission means that congestion management is no longer an internal matter. Of necessity, congestion management involves a system operator, transmission owners (if different from the system operator), power producers, and load-serving entities.

The separation of generation from transmission can lead to investment decisions in both sectors that are suboptimal from a broad societal perspective. For example, more than 8000 MW of new generating capacity plan to interconnect to the Palo Verde substation in Arizona (Emerson and Smith, 2001). But the existing transmission system can handle no more than 3360 MW of new generation. Even with the three new 500-kV lines proposed for this area, the maximum export capability will be only 6750 MW because of stability limits, well below the 8000 MW planned. To make the problem even worse, most of these new generators will obtain natural gas from the same pipeline. Thus, the outage of this pipeline could become the single largest contingency in Arizona, increasing greatly the amount of contingency reserves that must be maintained.

Finally, the amount and complexity of wholesale electricity commerce is much greater than it was a few years ago. Transactions today can span several control areas, and ownership of the power may change hands several times between the point of injection (the generator that produces the power) and the point of withdrawal (the load that consumes the power). This complexity makes it difficult for system operators to know the details of transmission flows and even more difficult to project what these flows might be like in future years.

Review of Recent Plans

Independent System Operators (ISOs) and utilities are developing transmission-planning processes to accommodate the needs of a rapidly evolving and increasingly fragmented electricity industry. This section briefly reviews several plans recently issued by ISOs and other regional entities.

The Electric Reliability Council of Texas (ERCOT) (2001) plan discusses historical and projected generation and load by region within ERCOT, including a range of projections. These projections form the basis for an identification of existing and likely future transmission constraints within the Interconnection and of an assessment of the need for additional transmission. The ERCOT report includes a discussion of existing transmission capacity and expansion possibilities for each of the three ERCOT subregions.

Overall, the ERCOT plan identifies six major transmission constraints (generally thermal limits, but sometimes stability limits). The plan also identifies several projects intended to mitigate these constraints. These projects include several 345-kV lines (both new lines and additional circuits on existing towers), a static compensator (to provide dynamic reactive-power support), and capacitors (to provide static reactive-power support). In addition, the transmission owners proposed several projects, which ERCOT recommended for construction.

One indication of the success of the ERCOT transmission-planning process is the number of transmission projects recently completed or under construction. ERCOT has several transmission advantages over other regions, including regulation by a single entity (the Texas PUC), a state government that supports additional transmission, and a regulatory system that gives transmission owners a reasonable assurance that their capital investments will be recovered. Of the seven projects considered critical during the past few years, one was completed in 2000, five are on schedule to be completed by the end of 2002, and one is undergoing further evaluation (Texas Public Utility Commission 2001).

The goals of the Mid-Continent Area Power Pool (MAPP) (2000) plan are to ensure that the transmission system can "reliably serve the load indigenous to the MAPP region,... provide sufficient transfer capability to reliably accommodate firm transfers of power among areas within MAPP and between MAPP and adjacent reliability regions, and provide an indication of transmission costs for enhancing transfer capability and relative costs for alternative locations of new generation." The MAPP process is bottom up, with plans developed by individual transmission owners, then integrated for each of the five subregions, and then integrated again at the MAPP level. In addition, considerable analysis is done for the MAPP region as a whole, primarily to analyze projects that span more than one subregion. The MAPP review ensures that projects proposed in one subregion will not adversely affect the electrical system in other subregions. Although MAPP planning still relies heavily on the individual utilities, the regional planning process is beginning to significantly influence the individual expansion plans.

The MAPP plan uses information on transmission service requests that were refused along with data on transmission curtailments to help in the analysis of "desired market use of the regional and inter-regional transmission system." These data "provided strong evidence to indicate that transmission constraints to the east of MAPP significantly hampered electrical sales" (Mazur, 1999).

The ISO New England (2001) plan breaks new analytical ground. This plan explicitly analyzed the potential benefits of new transmission from reductions in congestion through what the ISO calls its Projected Congestion Cost Assessment, "which, through modeling, determined the economic costs associated with transfer limits between regions and separately analyzed the New England system on a bus by bus basis for transmission constraints." As the report notes, "Significant transmission congestion will exist from an economic viewpoint, primarily between ME/NH [Maine and New

Hampshire] and Boston, SEMA-RI [Southeast Massachusetts] and both Boston and SWCT [Southwest Connecticut]. Estimates of New England congestion range between approximately \$200-\$600 million per year during the study period, depending on the assumptions utilized."

The New England analysis also considered the effects of market power on congestion costs, which could have enormous effects on the benefits associated with new transmission facilities. Because analysis of strategic market behavior is difficult, the New England analysis used a simple approach: it increased the bid prices for all generators by 10 or 25% above their marginal operating costs.[2] This approach may underestimate the benefits of transmission in reducing the ability of generators to exercise market power.

Figure 3 is a summary of some of the congestion-analysis results developed by ISO New England. The graph shows how sensitive these estimates are to different assumptions. And this is just a subset of the cases ISO NE examined; the 5- year congestion costs for the full set of cases ranged from about \$500 million to more than \$3 billion. Because this was just an initial assessment, it includes no estimates of the costs to build the transmission needed to relieve congestion in the region.

ISO New England divided the region into 13 subareas based primarily on transmission characteristics and constraints: "The subareas have been defined solely based on transmission interfaces that are relevant to both reliability and congestion concerns." These subareas do not necessarily conform to political or utility boundaries.

---

[2]The traditional assumption in production-costing models that generators bid their marginal costs is almost surely incorrect. On the other hand, appropriately simulating bidder behavior, with and without new transmission that expands the scope of regional markets and reduces congestion, is very difficult.

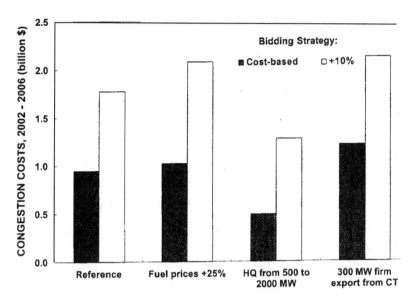

Figure 3. Congestion costs in New England under different assumptions about fuel prices, Hydro Quebec imports, and exports from Connecticut, as well as the bidding behavior of generators.

National Grid USA (2000) owns some of the transmission assets in New England. Its report is less a detailed plan for New England and more an overview of likely transmission needs in the future. The report examined the period 2001 through 2005 in terms of demand projections, generation, the relationship of generation to demand, transmission-system topology (major zones and interfaces), transmission performance (system power flows), capability (transfer limits and congestion costs), and transmission-system opportunities.

The chapter on opportunities is especially interesting because it shows where within New England new generator interconnections "would alleviate or exacerbate congestion on the transmission system." As Figure 4 shows, locating generators in Boston or southwest Connecticut would relieve congestion, whereas locating generators in

Maine, northern Vermont or New Hampshire, Rhode Island, or southeastern Massachusetts would worsen congestion. Information like that shown in Figure 4 should help guide market decisions on new generation and load-management programs, as well as possible merchant-transmission projects. Provision of information on the current and expected future state of the transmission system and the costs of using that system could reduce what North American Electric Reliability Council (NERC) (2001) sees as "inefficient transmission expansion [caused by the] uncoordinated siting of generation and the development of transmission projects."

The initial assessment conducted by American Transmission Company (2001) divided its region into five subareas. Like ISO New England, ATC defined these zones on the

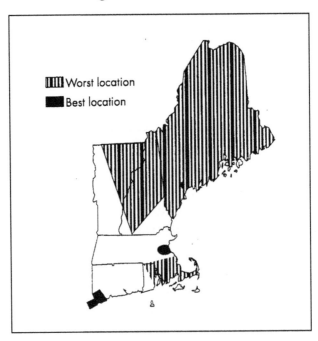

Figure 4. National Grid USA's assessment of the best and worst locations within New England to locate new generating units.

basis of transmission "system topology, load characteristics, load density and existing generation." ATC plans to revise the boundaries of these zones if and when bulk-power flows and conditions change. Its initial plan presents several proposals for transmission additions within each zone for 2002, 2003, 2004, and between 2005 and 2010, based on loadflow simulations conducted for 2002, 2005, and 2010.

The Western Governors' Association (2001) issued a conceptual plan for the Western Systems Coordinating Council. The plan is conceptual in that it looked at broad regional needs and not at local transmission needs. The report noted important limitations in current transmission plans and the associated planning processes:

> The current transmission planning process is fragmented, based on utilities' individual forecasts of needs and specific interconnection requests from new generation.

> At best, coordination occurs on a subregional basis. The current planing process is reactive, rather than forward looking. There is a wide gap between evolving merchant needs on the resource side (regional) and existing grid plans (local or sub-regional) on the transmission side. Planning assumptions are based primarily on local traditional resources and give little consideration to remote and nonconventional resources.

This western analysis considered two generation scenarios to the year 2010. One involves gas-fired generation built close to load centers and the other includes coal, wind, geothermal, and other generation located in remote areas. In the first scenario, little new transmission is needed between 2004 and 2010. In the second case, transmission investments of $8 to $12 billion are needed to support 23 GW of new remotely located generation.[3]

Because the Bonneville Power Administration has built no major transmission facilities since 1987, it has a substantial backlog that it is now addressing (VanZandt, 2001). Experts from eight electric utilities in the Pacific Northwest reviewed the first nine projects that BPA proposed, at a total cost of $615 million (Infrastructure Technical Review Committee, 2001). This largely qualitative review examined, for each of the nine projects, the limiting outages to be addressed by the project, the expected local and regional benefits from the project, various risks associated with the project, a project description, and alternative transmission projects that could address the limiting outages. The review also includes an appendix on risk and uncertainty that outlines the kinds of risks facing new transmission projects, including adequacy requirements, congestion relief, changes in electric-industry structure, and over- vs under-building.

Some recent plans are more limited in scope than the ones discussed above. Often, the plans do not fully integrate planning for reliability with planning for commerce. Because some entities have received so many generator- interconnection requests, their plans are dominated by the specific projects required to connect these new generators to the grid. Correspondingly, the plans do not anticipate possible problems that might occur in the future as a consequence of load

---

[3]This works out to a transmission investment of more than $400 per new kW of remote generation, a very high cost. If new coal and wind generation costs about $1000/kW, the supporting transmission would add 40% to the initial cost. By comparison, the new transmission planned for the Pennsylvania-New Jersy-Maryland Interconnection (PMJ, 2001b) region ($720 million to connect 27,500 MW of new generation) is expected to cost only $26 per new kW of generation. Part of this cost difference occurs because the distances between generation and load centers are much greater in the west than in the mid-Atlantic region.

growth; generator retirements; other new generators being built within the control area; or additional bulk-power transactions into, out of, or through the control area. In particular, these plans generally do not provide sufficient guidance to market participants on desirable locations for new generation, load-reduction programs, or merchant transmission. These plans are more reactive than proactive, in part because transmission planners do not have enough time to develop plans that look out several years and offer guidance on where to locate new generators. Instead, the planners are often overwhelmed with requests for new generation interconnections. The Bonneville Power Administration (BPA, 2001) wrote:

> BPA has received requests for transmission integration studies for more than 13,000 megawatts of new generating capacity at sites around the Northwest. More are pouring through the door. In just the last two weeks, BPA has received eight formal requests for studies on integrating new combustion turbines totaling 3,850 MW. … The Transmission Business Line is informing developers that it will take at least nine to 12 months to complete the required studies.[4]

As of March 2001, PJM had received more than 250 generator-interconnection requests, organized into seven queues. The first five queues include 40 GW of new generation to be completed between 2001 and 2004, enough to add more than two-thirds to PJM's current generating capacity

---

[4]The Tennessee Valley Authority faces a similar situation. It has received applications from independent power producers for 90,000 MW of new generation, more than three times the amount of existing generation (Whitehead, 2001). TVA would need an extra 50 system planners to clear the backlog of interconnection studies associated with all these new generators.

(Figure 5). Similarly, ISO New England had, as of Spring 2001, a queue with 40 GW of new generation, far more than the region's peak demand of 23 GW.

Perhaps because of the many interconnection requests PJM has received, its plan, although massive in length and detail, appears to lack any overall purpose. The plan includes two baseline assessments, the first of which analyzes compliance with regional reliability standards from 2001 through 2006 assuming no new generating units are built. The second baseline plan examines, in a similar fashion, the years 2002 through 2007 assuming all new generation in Queue A[5] is built. Separately, PJM presents all the interconnection studies associated with the new projects in Queues A, B, and C. In August 2000, the PJM Board approved the Queue A construction projects, with an estimated cost of $300 million; in June 2001, the Board approved the projects in Queues B and C, estimated to cost an additional $420 million.

This review of recent transmission plans shows tremendous variation. No single plan encompasses all the elements of a good transmission plan, as discussed in the section on Proposed Planning Process. Several factors explain the lack of key elements in many plans: (1) the dramatic changes in the U.S. electricity industry raise new issues for transmission planning, (2) the data and analytical tools to address these new issues have not yet been developed, (3) the ISOs are new entities that are still expanding their staffs, (4) the au-

---

[5]PJM sorts generator interconnection requests into queues depending on when the request was formally made. The August 2001 PJM plan includes Queues A, B, and C, with a total of 27,500 MW of new generation. Although the use of queues may be fair to generators, its application is controversial because it may increase overall electricity costs. For example, some merchant generation projects, although far down in the queue, might help solve transmission problems and, from a societal perspective, should be expedited.

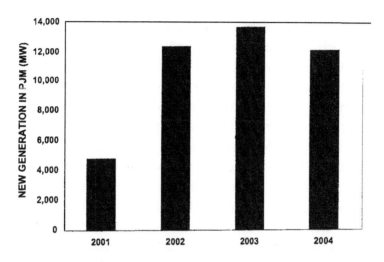

Figure 5. Planned generating capacity in the PJM area.

thority and responsibilities of the ISOs and other regional entities are not yet clear, and (5) the planning staffs are very busy responding to generator-interconnection requests. NERC (2001) recently noted that "these complex and rapidly evolving requirements are overwhelming the transmission planning process such that there is not enough time to develop optimal transmission plans."

Review of RTO Transmission Planning

The RTO filings of October 2000, required by FERC's Order 2000, pay little attention to Function 7 on transmission planning and expansion. The need to resolve other RTO issues—such as governance, regional scope and membership, and transmission-cost allocation and revenue requirements—dominated the prefiling deliberations. Perhaps because of these factors, FERC (1999) gave the RTOs three years after becoming operational to meet the requirements of Function 7.

The GridFlorida (2000) Planning Protocol calls for an "open and inclusive process" conducted by the RTO and supported by a Transmission Planning Committee that will provide "advice and input regarding the planning process" to the RTO. The protocol deals with regional planning; local planning; generation interconnection; data bases; standards for planning, design, and construction; transmission construction; and the role of reliability organizations and the Florida Public Service Commission in the planning process.

Although the GridFlorida proposal says much about the planning process, it contains few details on the substance of what a transmission plan should contain. While the protocol mentions "market solutions" it does not define the term and does not explain how they are to be identified, assessed, and implemented if found to be cost effective. Similarly, the protocol mentions "alternative solutions" but does not indicate what these alternatives might be, how they are to be compared with transmission solutions, and how they will be implemented.

The RTO West plan (Avista Corp. et al., 2000) "anticipates that RTO West's approach [to transmission planning] will evolve over time." The initial plan anticipates transmission expansion for two purposes: (1) "for reliability of service to load" and (2) "to relieve congestion." As noted elsewhere in this paper, distinguishing between reliability and commercial needs for new transmission is very difficult and perhaps a distraction. With respect to relief of congestion, RTO West anticipates a "market-driven expansion mechanism," which, in principle at least, should reduce the need for RTO West to develop its own plan in this area.

Attachment P (Description of RTO West Planning and Expansion) focuses on decision-making authority: who decides what facilities are to be built and who pays for these investments. The Attachment commits RTO West to develop:

(1) criteria to be applied by RTO West in determining the level of transfer capability that should be maintained from existing facilities, (2) transmission adequacy standards, (3) further definition of the market-driven mechanism [for transmission expansion], (4) the [new-transmission-cost] allocation procedure, including objective criteria, (5) interconnection standards, and (6) the details of the relationship/participation of RTO West with appropriate interconnection-wide and regional reliability organizations.

The Alliance RTO (American Electric Power Service Corp. et al. 2000) proposal is included in its Attachment H: Planning Protocol. The RTO is responsible for "coordinating" the planning rather than for doing the planning itself. (Some might question whether a "coordinated" plan is truly an integrated, regional plan or merely a collection of plans prepared by individual transmission owners.) The RTO's Reliability Planning Committee will be "the vehicle through which coordinated reliability planning activities will be conducted." RTO staff and representatives from each transmission owner and local distribution utility will be members of this committee, but not other market participants. This committee will be responsible for the planning models and data, reviewing and approving planning studies, determining the need for system expansion to meet reliability needs and transmission-service requests, participating in NERC and regional reliability processes, and coordinating transmission planning and expansion with other RTOs. The committee will produce a 10-year plan every year. The RTO's Planning Advisory Committee "will provide a forum for stakeholders and interested parties to have input in the planning process." With respect to transmission projects intended to reduce congestion, the Alliance RTO "will encourage market-driven operating and investment actions...."

The proposal from the New England Transmission Owners et al. (2001) builds on the experience with ISO New England. It envisions a binary RTO with a nonprofit ISO and a for-profit independent transmission company (ITC). The proposed three-phase planning process "combines the knowledge and objectivity of ISO-NE [ISO New England] with the strengths of an investor-owned business focused on transmission...." The process consists of the following steps:

- The ISO will lead a needs assessment, which will integrate data and projections on regional loads, generation (existing, planned retirements, and potential additions), transmission, and inter-control area transactions to forecast the region's needs for additional transmission. The needs assessment will be consistent with regional reliability planning standards, address congestion costs, and consider transmission-system performance.

- The ITC will develop a Regional Transmission Facilities Outlook, which will identify transmission alternatives that may be needed based on a range of plausible scenarios.

- Finally, the ISO will assess the ITC's Outlook and approve a regional plan. This assessment will consider other alternatives proposed by the ISO and stakeholders. The ISO review will provide "a check that the Outlook is not biased in favor of transmission solutions at the expense of generation or other market-based solutions." "The decision to proceed with [transmission projects] will be made by the market [participants] for market based proposals (including merchant transmission) and by the ITC for regulated transmission proposals."

This review of some RTO filings suggests that much work remains to be done by the RTOs to develop comprehensive and meaningful transmission-planning processes. Unfortunately, progress has been slow during the past several months. One RTO posted a progress report on its website in August 2001 that its "... planning and expansion principles are still under discussion...." Deciding on a specific transmission-planning approach is difficult in some regions because the participants cannot agree on whether transmission investments should be driven by the market participants or by reliability requirements. In the former case, generator owners, large customers, and private investors might pay for new facilities built as merchant projects, while in the latter case the transmission owners (and ultimately, all retail customers), in response to RTO plans, would pay for such projects through a centralized process.

Proposed Planning Process

As noted above, transmission planning today is much more complicated, and perversely, much more uncertain, than it was several years ago. Based on our review of several recent transmission plans, we offer a suggested RTO transmission-planning process (Figure 6), the results of which should include broad consensus on new transmission and nonwires projects that are needed and that get built in a timely and cost-effective fashion.

Several of the activities summarized below are covered in greater detail in the following section. Our proposed process begins with a clear identification of the purpose of the transmission plan (Step 1), followed by a comprehensive assessment of the current regional situation, encompassing

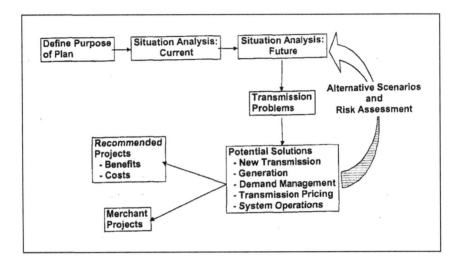

Figure 6. Outline of proposed RTO planning process.

both operations and markets (Step 2). This situation analysis provides a firm basis for discussing future conditions, problems, and potential solutions. Steps 3 and 4 involve projections of likely conditions several years into the future and an identification of transmission problems that might occur under these postulated future conditions. Steps 5 and 6 assess various transmission and nontransmission alternatives that might solve the problems identified in Step 4. Finally, Step 7 summarizes the results of the analyses conducted in the prior steps and recommends specific projects to address the transmission problems discussed in Step 4.

Step 1. What is the purpose of this transmission plan?[6] Who developed it? In response to what requirements? How were various interest groups (e.g., generators, transmission owners, load-serving entities, distribution utilities,

---

[6]These purposes could include maintenance of reliability, promotion of competitive electricity markets, support for development of new generation, promotion of economic growth, creation of new jobs, and so on.

retail customers, and state regulators) represented in the development and review of this plan? How does the plan reflect the market design in that region (e.g., the number and types of markets for energy, installed capacity, ancillary services, and congestion)? How were the practical limitations of siting and project financing addressed in the plan (e.g., did the planning process consider nontechnical as well as technical issues, who will pay for transmission projects)? To what extent is the plan intended to motivate market solutions to transmission problems?

Step 2. Describe the current situation, covering bulk-power operations (both generation and transmission), wholesale markets, and transmission pricing. What problems (e.g., reliability, congestion, losses, generator market power), if any, occur that are caused by limitations in the transmission system? What transmission projects are under construction or planned for completion within the next few years to address these problems? What are the estimated costs and benefits of these projects, individually and in aggregate? What entities are expected to benefit and to pay for these projects? Explain the computer models used to analyze transmission conditions and the limitations of these models (analytical approximations).

Step 3. Describe the bulk-power system as it is expected to exist in the future (e.g., five and ten years). What are the levels, patterns, and locations of loads? Describe the region's fleet of generating units, including locations, capacity, and operating costs (or bid prices). What are the likely effects of new generation facilities on interconnection requests, the overall transmission system, and the costs of new transmission construction? What transmission-pricing methods might be used to recover the costs of capital, losses, and congestion? Describe the transmission flows within the region as well as the flows that occur into, out of, and through the region. Given the many uncertainties that affect future fuel prices, loads, generation, transmission and its pricing, and market rules, create various scenarios or sensitivities that can be used subsequently to analyze potential problems and transmission improvements (Steps 4 and 5).[7]

Step 4. What transmission problems, both reliability and commercial, are likely to exist given the future conditions (scenarios) developed in Step 3?[8] What other problems might exist for which transmission could be applied (e.g., generator market power caused by a restricted geographical scope of wholesale markets, limited fuel diversity caused by insufficient transmission facilities to remote locations with fuel, such as coal and wind)?

Step 5. What transmission facilities might be added to the current system to address the problems identified in Step 4? What effects would these facilities have on compliance with reliability standards, commercial transactions, losses, and overall regional electricity costs (generation plus transmission)? Can recent technological advances in transmission equipment and software be applied? Do they capture potential economies of scale

---

[7]The results of Steps 2 and 3 should be sufficiently detailed that other parties can assess for themselves market solutions to solve these problems (e.g., those discussed in Step 6).

associated with building (ahead of need) larger lines than currently needed? Do these proposals address the potential for generators to exercise market power in wholesale electricity markets?[9] What are the likely capital costs of these transmission additions? How do the costs and benefits of individual projects, as well as groups of projects, compare with each other? Can any of these transmission projects be built on a merchant (i.e., for profit and unregulated) basis? What kinds of risk assessment were conducted in developing recommendations on these new transmission projects?[10] How were these risks addressed in the plan, including the risks of over- vs underbuilding transmission?[11] Should certain transmission facilities be built to guide current and future decisions on the locations of new generating units and the locations and types of demand-management programs; that is, should transmission planning be proactive rather than reactive?

---

[8]These problems could appear as real-time congestion, denial of requests for service, or curtailment of approved transactions. They could also include operational difficulties caused by aging and obsolete equipment that should be replaced to reduce forced and maintenance outages or increase functionality.

[9]It may be very difficult analytically to estimate the kinds of strategic bidding behavior that might occur. Such behavior will be a strong function of the RTO operating and market rules as well as the physical infrastructure (amounts and locations of generation, transmission, and load).

[10]Uncertainties are much greater than in the past. Today, they include load shape and levels, generator locations (new construction and retirements), market operations, market prices for energy and ancillary services, transmission pricing (including locational pricing for losses and congestion), patterns and levels of commercial transactions, weather, fuel price volatility, and new generation and transmission technologies.

[11]For example, consider the risks associated with cost recovery for a new transmission line needed to connect a new generator to the grid. This risk could be eliminated by requiring the generation owner to pay the capital costs up front rather than through rates over a 20-year cost-recovery period.

Step 6. What nontransmission alternatives (including suitably located generation and price-responsive load programs as well as alternative transmission-pricing schemes[12]) might be deployed to address the problems identified in Step 4? These alternatives could also include changes in system-operations, such as remedial-action schemes. To what extent can these generation, demand-side, and pricing alternatives address the problems for which the transmission facilities suggested in Step 5 were proposed? What are the expected costs to the transmission system of these nontransmission alternatives (which may not reflect the total costs of these generators and/or demand-management programs)? Based on the differences in characteristics and the differences in costs and benefits, recommend either transmission or nontransmission solutions to the problems identified in Step 4. Where no solutions are offered, indicate why. (Presumably, the expected status quo should continue if the costs of solving a problem exceed the benefits of doing so.)

Step 7. Based on the foregoing analyses, recommend transmission pricing, generation-location decisions, demand-management programs, and construction of new transmission facilities. If market participants do not propose the solutions analyzed in Steps 5 and 6, recommend those transmission facilities (from Step 6) that should be built under traditional regulation. Summarize the benefits and costs of these proposed projects, both singly and in aggregate. Can the projects ultimately be approved and built in a timely fashion? Can they be fi-

---

[12]Such pricing schemes should encompass access charges as well as charges for congestion and losses.

nanced? Will these projects be undertaken by market participants acting in their own interest, or must the RTO require their construction and ensure that customers at large pay for them?

Table 2, based on this 7-step process, identifies key ingredients of a successful transmission planning process and plan.

Table 2.
Checklist of important characteristics of a transmission plan

- Public involvement throughout planning process

- Broad range of alternatives considered, including suitably located generation and demand-management programs, new transmission technologies, and various transmission-pricing methods

- Effects of transmission on generator market power

- Effects of transmission on compliance with reliability standards, both planning and operating

- Effects of transmission on congestion costs

- Comprehensive risk assessment of transmission plan(s)

- Proactive, rather than reactive, transmission plan (consideration of needs for increased throughput and locational guidance for new resources, not just responses to generator-interconnection requests)

- Development of a practical and robust, rather than a theoretically optimized, transmission plan

- Support for projects built through competitive-market mechanisms

- Timely completion of the plan

KEY TRANSMISSION PLANNING ISSUES

Planning Criteria: Reliability and Commerce

Traditionally, vertically integrated utilities planned their transmission systems to: (1) meet North American Electric Reliability Council (NERC) and regional-reliability-council reliability requirements and (2) ensure that the outputs from the utility's generation could be transported to the utility's customers. (Utilities sometimes built transmission lines for economic reasons; for example, to provide access to cheaper power in a neighboring system or to export surplus power.) Today, transmission systems are called on to do much more. They must serve dynamic and rapidly expanding markets in which the flows of power into, out of, and through a particular region vary substantially over time. As a consequence, transmission planners may need to look beyond the NERC Planning Standards in assessing alternative transmission projects and also consider enabling competition to occur over large geographic regions (NERC 1997). A market-focused approach might seek to reduce the number of times transmission-service requests are denied and generation must be redispatched to avoid congestion. Where congestion (locational) pricing is used, this goal is met by reducing congestion costs (discussed below). Congestion pricing might reduce the distinction between reliability and commerce by explicitly pricing reliability.

Many industry experts believe that the distinction between reliability and commerce in transmission planning is increasingly irrelevant. Reliability problems (e.g., a line that would become overloaded during a contingency) are also commercial problems that affect different market participants differently (e.g., flows are reduced on the line in question, which means that the output from cheap generators must be reduced and the output from expensive generators must be increased). Conversely, certain commercially desirable flows may be restricted because of reliability problems that would otherwise occur. Equally important, these people believe that transmission serves a vital enabling function, permitting the purchase and sale of energy and capacity across large regions and, in the process, reducing problems associated with generator market power.

Some experts believe that the distinction between reliability and commerce is important. Not all reliability problems have commercial implications, they noted. Some local problems (e.g., low voltages close to load centers) are related more to reliability than to commerce. The solution to such reliability problems might be the addition of capacitors to serve local loads regardless of whether the generation source is near or far. The distinction may be important in determining who pays for the project, with reliability projects paid for by all grid users but commercial projects paid for only by those transmission customers who benefit from the project. Of course, deciding who does and does not benefit from a project can be difficult and contentious. The Pennsylvania-New Jersey-Maryland Interconnection (PJM) (2001a) baseline plan focuses on reliability: "Transmission constraints on market dispatch are economic constraints. Economic constraints are not considered violations of reliability criteria as long as the system can be adjusted to remain within reliability limits on a pre-contingency basis."

Economies of Scale

It is generally cheaper per megawatt of capacity to build larger transmission lines (Table 3). For example, the cost per MW-mile of a 500-kV transmission line is about half that of a 230-kV line. Higher-voltage lines also require less land per MW-mile than do lower-voltage lines (right side of Table 3). A 500-kV line requires less than half the land per MW-mile of a 230-kV line.

Both of these factors argue for overbuilding lines rather than trying to size lines to exactly match current and short-term forecast needs. (Overbuilding includes the use of larger conductors, construction of larger towers that can carry more than one set of circuits, and the use of higher-voltage lines.) Overbuilding a line now will (1) reduce long-term costs by avoiding the much higher costs of building two smaller lines and (2) reduce the delays and opposition associated with transmission-line siting by eliminating these costs for the now unneeded second line.

Table 3. Typical costs, thermal capacities, and corridor widths of transmission lines

Voltage (kV)	Capital cost[a] (thousand $/mile)	Capacity (MW)	Cost (million $/GW-mile)	Width[b] (feet)
230	480	350	1.37	100
345	900	900	1.00	125
500	1200	2000	0.60	175
765	1800	4000	0.45	200

[a]These estimates are from Seppa (1999) and include the costs of land, towers, poles, and conductors. We increased these estimates by 20% to account for the costs of substations and related equipment.
[b]These estimates are from Pasternack (2001).

On the other hand, the lumpiness of transmission investments (e.g., one can build a 345-kV line or a 500- kV line but not a 410-kV line) can complicate decisions on what to build and when. Also, a large transmission line may impose more of a reliability burden on the system than do several smaller lines. Indeed, if a new, large line becomes the largest single contingency, contingency-reserve requirements might increase in the region. And, opposition might be greater to a 500-kV line than to a 345-kV line because the former line has taller towers and requires more land.

Congestion Costs

Decisions on whether to build new transmission are complicated by uncertainties over the future costs of congestion. (To some extent, the prices of firm transmission rights show how the market values certain transmission paths.) These uncertainties relate to load growth, the price responsiveness of load, fuel costs and therefore electricity prices, additions and retirements of generating capacity and the locations of those generators, the exercise of market power by some generators, and transmission pricing. The ISO New England (2001) analysis, summarized in Figure 3, shows this complexity very well. Analysis conducted for the New York ISO showed that the large number of proposed generating projects in or near New York City and Long Island "would reduce the level of congestion observed on the...bulk power system, with the biggest congestion decreases occurring in New York City and on Long Island" (Sanford, Banunarayanan, and Wirgau, 2001).

We developed a simple hypothetical example to explore these issues and their complexities and interactions. This example involves two regions, A and B, separated by 200 miles. Region A contains 31 GW of generating capacity and no load. Region B contains 32 GW of generating capacity and 50 GW of load. Both regions contain a wide range of generating capacity, with running costs (or bids) that vary from zero to almost $160/MWh. The load in Region B ranges from 20 to 50 GW, with a load factor of 63%.

We calculated the cost of congestion as the difference between (1) the cost of generation (including generators in both regions) to serve the load in Region B when transmission capacity between the two regions is limited, and (2) the cost of generation when transmission capacity between the two regions is infinite. The generation costs in both cases are calculated for every hour of the year.

Figure 7 shows the cost of congestion as a function of the transmission capacity connecting the two regions. With 21 GW of transmission capacity (the baseline in this example), electricity consumers in Region B pay $87 million a year because of congestion. As the amount of transmission capacity increases, the cost of congestion declines because the number of hours that congestion occurs and the price differences between A and B decline. However, as shown in Figure 7, this decline is highly nonlinear, with each increment of transmission capacity providing less and less economic benefit. Expanding transmission capacity from 20 to 21 GW lowers the cost of congestion $99 million/year, expanding capacity from 21 to 22 GW saves $44 million, and

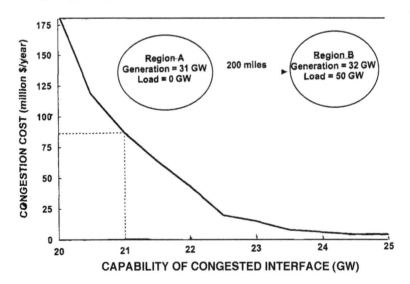

Figure 7. The annual cost of congestion as a function of transmission capability between hypothetical regions A and B.

expanding capacity from 22 to 23 GW cuts costs by only $29 million.

The relationship between the benefits of adding transmission capacity between A and B to reduce congestion costs and the costs of doing so are highly nonlinear because of (1) nonlinearities in congestion costs, (2) economies of scale in transmission investments, and (3) the lumpiness of transmission investments. For this example, if the goal is to increase capacity by 0.5 GW, it makes sense to build either two 230-kV lines or one 345-KV line, but not a 500-kV line. On the other hand, it is most cost effective to use 500-kV lines when expanding capacity by 1 GW or more. Indeed, the benefit/cost ratio for 230-kV lines increases in going from an addition of 0.5 to 1.0 GW, but then declines as more capacity is added. On the other hand, the benefit/cost ratio is more than 2 for the addition of a 500-kV line to expand capacity by 1.5 or 2.0 GW.

What happens to these costs and benefits if additional generating capacity is built in Region B, close to the load center? Adding 0.5 GW of capacity with a running cost of $30/MWh reduces congestion costs by TXCongestion $19 million/year. Adding 2 GW of such capacity reduces congestion costs by $59 million/year. If the new generating capacity added to Region B had a running cost of $57/MWh, its congestion-reduction benefits would be only $14 and $35 million/year for 0.5- and 2-GW additions, respectively. These benefits are about two-thirds of those that would occur with new capacity at $30/MWh. Clearly, building new generation in Region B would undermine the economics of adding transmission capacity between regions A and B.

The congestion-reduction benefits of each additional MW of generating capacity are less than the benefits of earlier additions. This effect is especially pronounced as the bid prices of the new units increase. For the more expensive of the two units there is no benefit from adding more than 1.5 GW of generating capacity in Region B because other generators are less expensive. Once again, the results are highly nonlinear.

If loads grow at 2% a year, the annual cost of congestion (assuming no additions to either generating or transmission capacity) increases from $87 million in the initial year to $125, $162, and $250 million in the second, third, and fourth years. Such increases in load make transmission investments substantially more cost-effective. If loads respond to prices, such that loads are higher at low prices and lower at high prices, congestion costs would be reduced. In this example, as the price elasticity of demand increases from 0 to 0.02, 0.04, and 0.08, congestion costs are reduced from $87 million to $48, $25, and $7 million a year. For the ranges in load growth and price elasticity considered here, congestion costs vary from $7 to $250 million a year when the amount of transmission capacity between the two regions is 21 GW. Making decisions on how much money to invest in equipment with lifetimes of several decades is difficult in the face of such uncertainties about future load growth; customer response to dynamic pricing; and the amounts, locations, and running costs of new generating units.

The discussion so far has focused on the benefits of reducing congestion. But not all market participants benefit when additional transmission is built to relieve congestion. In particular, loads on the low-cost side of the constraint and generators on the high-cost side of the constraint lose money when congestion is reduced. For example, a generator in Region B with a bid price of $42/MWh would earn $6.9/kW-year when the transmission capacity between regions A and B is 20 GW. Expanding transmission capacity to 21 or 22 GW would reduce that generator's earnings to $4.6 and $3.7/kW-year, reductions of 33% and 46%, respectively. Such large prospective losses would likely engender substantial opposition to efforts to reduce congestion. (If Region A had loads that enjoyed the benefits of Region A's low-cost generation, those loads would also oppose efforts to reduce congestion.)

Finally, investors considering additional generation in Region B may worry that future construction of a new transmission line between A and B would undercut the value of their new generation.

Generation and Load Alternatives

The Department of Energy Task Force on Electric System Reliability (1998) recommended that RTOs "ensure that customers have access to alternatives to transmission investment including distributed generation and demand-side management to address reliability concerns and that the marketplace and the [RTO's] standards and processes enable rational choices between these alternatives."

Transmission planners can encourage nontransmission alternatives in two ways. The simplest method is to provide transmission customers with information on current and likely future congestion costs. Such information—coupled with locational pricing for congestion and losses—on the costs and benefits of locating loads and generation in different places could motivate developers of new generation to pick locations where energy costs are high, thereby reducing congestion costs. Similarly, such information could motivate loadserving entities to offer load-reduction programs to their customers in those areas where energy prices are high because of congestion. For example, the National Grid USA (2000) transmission plan included a map of New England (Figure 4) showing areas where new generation would worsen congestion and areas where new generation would reduce congestion. An alternative approach to the provision of information only is to pay for nontransmission alternatives. With this approach, the RTO would first prepare a transmission plan. This plan would likely include one or more major transmission projects (new lines and/or substations). Next, the RTO would issue a request for proposals for alternatives and then review the proposals to see if they were less expensive than the original transmission project and provided the same or better reliability and commercial benefits that the transmission project would. Ultimately, the least-cost solution to the identified transmission problem would be acquired by the RTO and recovered through transmission rates.

Appropriately comparing transmission to load or generation, however, is difficult because they differ in construction leadtimes, project lifetimes, availability, capital and operating costs, market type, and technical applicability:

- Lifetimes—Transmission investments are long-lived (30 to 50 years). Generators typically have shorter lifetimes, and load-management projects may have much shorter lifetimes (e.g., if a building is extensively remodeled, the load-management systems may be removed and replaced with alternative systems for lighting, heating, cooling, and ventilation). The longer lifetimes of transmission projects enhance confidence in their ability to provide the needed service for many years; however they reduce flexibility to respond to changed circumstances in the future.

- Availability—Transmission equipment typically has very high availability factors, much higher than those for either generation or load.

- Capital and operating costs—Although the capital costs of transmission can be high, operating costs are very low. The operating costs for generators are high and depend strongly on uncertain future fuel prices. The trade-off here is between high sunk costs (once the transmission project is completed) against uncertain operating costs for generation and load management.

- Type of market—The returns on transmission investments are regulated, today primarily at the state level and in the future primarily by FERC. The profitability of generation investments, on the other hand, is determined largely by competitive markets. Comparing costs (e.g., economic lifetimes and rates of return) between regulated and competitive markets is difficult.

- Technical applicability—Nonwires resources cannot always solve the problems at which the transmission investment is aimed (e.g., transient stability or the need to replace aging or obsolete transmission equipment). Also, connection of the resource to the grid may impose new costs on the system (e.g., if system-protection schemes must be upgraded).

The difference in lifetimes between the transmission project and its alternatives raises the issue of whether the alternatives should be assessed against the cost of deferring the transmission project for several years or against the full cost of displacing (eliminating the need for) the transmission project. If the transmission project will likely be needed in any case, although at a later date, the deferral approach makes sense.

Although the concept of encouraging competition between transmission investments and generation and load alternatives is appealing, implementation can be difficult. The Tri-Valley project, proposed by Pacific Gas & Electric in northern California, illustrates these difficulties. The project involves the construction of new 230-kV transmission lines, construction of new 230/21-kV substations, and upgrading of a substation to 230-kV service. The California ISO issued a request for "cost effective and reliable alternatives... from generation and/or load alternatives to the proposed PG&E transmission project" (Winter and Fluckiger, 2000). Alternatives were required to be available between the hours of 8 am and 1 am for up to 500 hours between April 1 and October 31 each year from 2001 through 2005. The ISO sought call options on about 175 MW. The request was issued in January 2000 with responses due in late March. The ISO received four proposals, all of which it subsequently rejected.

The ISO rejected all four bids because they failed one or more of the evaluation criteria, which involved satisfaction of the ISO's reliability criteria, commencement date, operating characteristics, ability to provide the proposed services, cost, safety, impacts on markets (in particular, effects on generator market power), and environmental implications. The key reason the bids were rejected is they were substantially more expensive than the transmission project. Also, the transmission project was expected to provide more capacity to the system than the generation and load-management projects.

A year later, when faced with a similar situation, the ISO decided against issuing a competitive solicitation. In this case, the ISO approved construction of the San Diego Gas & Electric Valley-Rainbow transmission project (Detmers, Perez, and Greenleaf 2001). In part because of the electricity crisis California faced, the ISO decided that this project should be considered part of a "broad strategy by the state of California to put into place a robust transmission system to support reliable service to consumers." The benefits of this

500- kV transmission project would not be realized by generation or load-management alternatives. The proposed transmission line would permit generation from other parts of California, Arizona, and New Mexico to be moved to the San Diego area. The project would also permit new generators being located near San Diego to reach distant markets. Finally, the project would provide local reliability benefits that otherwise would have to be purchased through reliability-must-run contracts. These reliability benefits would occur because the transmission project "integrates San Diego with the rest of the Western Interconnection, providing significant access to a wide variety of resources rather than being limited to the local area resources and the common concerns that they share, such as adequacy of gas supply."

The limited analysis conducted to date seems to argue against widespread use of suitably located generation and load management as alternatives to some new transmission projects. However, these analyses were conducted primarily by transmission engineers who are more comfortable with transmission and understand transmission better than they do its alternatives. Also, the continued opposition to construction of new transmission facilities requires the electricity industry to look long and hard at possibly viable alternatives.

New Technologies

Superconductivity, power electronics, information systems, and other new technologies could revolutionize transmission and make it easier to expand the system through merchant, rather than regulated, projects. According to Howe (2001), "Recent advances in materials science offer the prospect of another industry paradigm: one based on robust facilities-based competition in network services, without the environmental and land-use impacts of traditional 'big iron' solutions." Some of these advances include:

- Superconducting Magnetic Energy Storage—High-speed magnetic-energy-storage devices that are strategically located in a transmission grid to damp out disturbances. These systems include a cryogenically cooled storage magnet, advanced line-monitoring equipment to detect voltage deviations, and inverters that can rapidly (within a second) inject the appropriate combination of real and reactive power to counteract voltage problems. By correcting for potential stability problems, these systems permit the operation of transmission lines at capacities much closer to their thermal limits than would otherwise be possible.

- High-Temperature Superconducting Cable—Can carry five times as much power as copper wires with the same dimensions. Although initially applicable to underground distribution systems in dense urban areas, eventually this technology may be used for medium- and high-voltage underground transmission lines. The use of these cables would greatly reduce the land required for transmission lines in urban areas and lessen aesthetic impacts and public opposition.

- Flexible AC Transmission System (FACTS) devices—A variety of power-electronic devices used to improve control and stability of the transmission grid. These systems respond quickly and precisely. They can control the flow of real and reactive power directly or they can inject or absorb real and reactive power into the grid. These characteristics provide both steady-state and dynamic benefits. Direct power-flow control makes the devices useful for eliminating loop flows. The very fast response makes the devices useful for improving system stability. Both characteristics permit the system to be operated closer to its thermal limits. FACTS devices include static var compensators, which provide a dynamic source of reactive power; thyristor-controlled series capacitors, which provide variable transmission-line compensation (effectively "shortening" the line length and reducing stability problems); synchronous static compensators, which provide a dynamic source of reactive power; and universal power-flow controllers, which control both real- and reactive-power flows.

- High-voltage DC (HVDC) systems—HVDC lines have several advantages over AC transmission lines, including no limits on line length, which is useful for moving large amounts of power over long distances; reduced right-of-way because of their more compact design; precise control of power flows, eliminating loop flows; and fast control of real- and reactivepower to enhance system stability. The primary drawback of HVDC is the high cost of the converter stations (which convert power from AC to DC or vice versa) at each end of the line.

- HVDC Light—This new approach to HVDC uses integrated-gate bipolar transistor-based valves (instead of thyristor-based valves) in the converter stations. These new valves permit economical construction of lower-voltage lines, which greatly increases the range of applicability for DC lines; involves much more factory construction instead of on-site construction, which lowers capital costs; and provides better control of voltages and power flows. HVDC light lines have recently been built in Australia and Denmark, and others have been proposed for the United States.

- Real-time ratings of transmission lines—Represent another use of advanced information technologies to expand the capability of existing systems (Seppa, 1999). Such systems measure the tension in transmission lines, ambient temperature and wind speed, or cable sag in real time; the results of these measurements are telemetered to the control center, which then adjusts the line rating according to actual temperatures and wind speeds.

In spite of their wonderful attributes and recent declines in their costs, these technologies are generally too expensive to warrant their widespread use today. (To date, they have been deployed in a few locations, primarily by utilities to improve the performance of their systems.) However, as the technologies are improved and demonstrated, their costs will likely continue to drop enough that they become cost effective. When that day arrives, transmission planning will be simpler, primarily because market participants (rather than regulators or system operators) will be able to decide whether to invest in these systems and will be able to retain their benefits (because some of these technologies use devices that permit direct control of power flows).

Merchant Transmission

The kinds of new technologies discussed above make it possible for unregulated, for-profit entities to build what are called merchant transmission projects. Under such circumstances, the need for centralized transmission planning is greatly reduced. Three such merchant projects have been proposed in the United States:

- TransEnergie US proposes to build a 330-MW, 26-mile submarine cable under Long Island Sound to connect

Connecticut and Long Island. FERC approved the project in June 2000, after which TransEnergie held an open-season subscription for the DC line's capacity.

- The Neptune Regional Transmission System, announced in May 2001, is a set of DC projects to link the northeastern U.S. with eastern Canada. All four phases involve submarine cables. The total project calls for 3600 MW of transfer capability from Canada to the U.S. FERC approved the project in July 2001.

- The TransAmerica Grid, proposed by Black & Veatch and Siemens AG, calls for construction of large mine-mouth coal plants in Wyoming and DC lines to connect this new generation with Chicago and Los Angeles. These transmission lines, about 1000 miles each, would cost $4.5 billion and would greatly expand the transfer capability between the eastern and western interconnections. All three of these projects are DC. As noted by Liles (2001):

"...the benefit of DC lies in the ability of the project's operator to control the flow of power on the line. What you put in is what you get out, net of resistive losses. Loop flow is not an issue. Contrast that with the existing AC network, in which power flows freely throughout the system according to the impedances of the lines.... Physically firm transmission capacity can be bought and sold on a DC line. For DC lines, the contract path is the actual path over which the power flows."

Such merchant projects are feasible only if the owner can obtain clear property rights to the transmission investment. According to Rotger and Felder (2001), such property rights require the use of "bid-based, security- constrained locational pricing for transmission services" as well as financial transmission rights. The PJM and New York ISOs have such systems in place.

Rotger and Felder (2001) propose a regulatory backstop in case competitive markets do not construct enough transmission to maintain reliability. Their vision of a backstop, however, is quite limited. It calls for the RTO to assess the need for new transmission to meet reliability requirements only, with no consideration of economic projects that might reduce costs to market participants. The RTO, having identified transmission projects needed for reliability, would then issue a request for proposals for such projects (Figure 8).

Although attractive in concept, no merchant transmission projects have yet been built in the United States. It is also unclear whether such projects are viable only where direct control is possible (e.g., with DC lines and other new technologies such as FACTS systems) or whether such

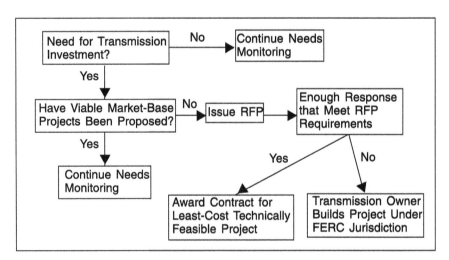

Figure 8. Proposed RTO backstop process to be used when competitive markets do not produce enough transmission expansion to meet reliability requirements.

projects are feasible for AC systems. If merchant projects are limited to those where control is possible, it is unclear whether such projects will play a major role in expanding North American transmission systems or will play more of a niche role.

Projections of New Generation and Load Growth

The deintegration of the traditional utility, which encompassed generation, transmission, distribution, and customer service in one entity, raises two important informational issues for transmission planning. First, from what sources will transmission planners obtain reliable information on the locations, types, capacities, and in-service dates of new generation? Second, what entity will be responsible for developing projections of future load growth?

Historically, utilities reported their plans for new generation to the Energy Information Administration (EIA) and NERC. Increasingly, however, new generation is being constructed by independent power producers. Although EIA collects data from such entities, long lags can occur between the time a company announces a new power plant and the time it shows up in the EIA system. The Electric Power Supply Association also collects data on power-plant construction plans. Because the Association does not provide details on the status of the project, it is hard to determine the probability that a project will get built and produce power. The probability of unit completion increases as the project moves from initial announcement to applications for siting and on to environmental permits, construction, and completion.

Analogous issues concern projections of future load growth. System operators (ISOs and, in the future, RTOs) monitor and record data on power flows down to the level of distribution substations. But, because of their focus on bulk-power flows and wholesale electricity markets, system operators are unlikely to have data on end-use demand by customer class. The competitive load-serving entities may have such information but are unlikely to want to make such information publicly available. The electricity industry needs to develop a system to collect relevant data on customer electricity-using equipment, load shapes, and load levels and to provide this information to transmission planners (as well as to other entities responsible for maintaining a healthy bulk-power system).

Recommendations

As the electricity industry continues its long and complicated transition to a fully competitive state, the requirements for transmission planning are changing and expanding. This paper outlined a proposed planning process that RTOs might adopt in a restructured electricity industry. However, most of the details for this process are not yet developed. Similarly, FERC's requirement in Order 2000 that "the RTO must have ultimate responsibility for both transmission planning and expansion within its region" is largely undefined. These gaps lead to several recommendations for the U.S. Department of Energy (DOE), FERC, and RTOs to consider:

- Provide technical assistance to ISOs, RTOs, regional reliability councils, federal power agencies, and other organizations to develop and demonstrate improved transmission-planning methods. Such methods would feature active public involvement throughout the planning process, comprehensive consideration of nonwires solutions to transmission problems, analysis of the benefits and costs of different solutions under a wide range of possible futures, and a focus on practical solutions that can be readily implemented. DOE could work with the planning staffs at various electricity-industry organizations to develop improved planning processes, analytical tools, and plans. DOE could then widely disseminate the results of these case studies (i.e., through publications, conferences, and workshops) so that others in the electricity industry can learn from these experiences.

- Assist FERC in the development of planning standards that FERC would then use in its review and approval of RTO transmission plans. This activity would add detail to the FERC Order 2000 requirement that RTOs be responsible for planning (Function 7). Based on the case studies described above, DOE could work with FERC staff to define what public involvement is required, what data RTOs must make available to market participants on the current and likely future states of the transmission system, what FERC means by "least cost" in its requirement that RTOs be responsible for transmission planning and expansion, and the extent to which planning should be proactive (i.e., guide future investments in new generation and demand-management programs), rather than only react to generator-interconnection requests and load growth. These standards should focus on performance (what is to be accomplished) and not be prescriptive, to permit flexibility within and among RTOs.

- The RTOs, acting under FERC requirements, could ensure that transmission planning and expansion fully comply with NERC and regional planning standards. Such compliance would ensure that transmission systems are adequate and meet reliability and commercial needs.

- The RTOs should identify the transmission-information needs of market participants (including generation developers, load-serving entities, transmission owners, and others) to guide their investment and operating decisions so they are consistent with current and likely future transmission conditions and costs. The information needs of interested stakeholders will vary considerably. Some participants will only want maps showing "good" and "bad" locations for new generation from the perspective of the transmission system, while other participants will want detailed load-flow studies that show voltages and flows throughout the system, under various on- and off-peak conditions. Periodically, such information should be made available to market participants.

- Study the potential role of merchant transmission. DOE, again working with RTOs and other market participants, could conduct a study to determine the extent to which merchant (nonregulated) transmission projects can meet future transmission needs. Among other topics, this study should examine the possibility of extending merchant transmission to AC projects, rather than the DC projects that are the focus of today's merchant transmission facilities. Another critical issue concerns the meaning of the RTO role as a "backstop" to market solutions. Under what conditions should the RTO build (or pay for) a project that is needed to solve transmission problems that market participants have not, acting on their own, chosen to solve? This study should also address the danger that merchant transmission will "cherry pick" the most profitable transmission projects, leaving the regulated entity (more accurately, transmission customers in general) to pay for the less cost-effective transmission projects that, nevertheless are required for reliability or to connect customers to the system.

Conclusions

Maintaining a healthy transmission system is vital for both reliability and commerce. Because electricity is essential to our modern society, public policy should ensure suitable expansion of the nation's transmission grids. Unfortunately, the historical record shows a clear and long-term decline in U.S. transmission adequacy (Hirst and Kirby 2001). Specifically, the amounts of new transmission added during the past two decades have consistently lagged growth in peak demand. To make matters worse, projections for the next five and ten years show continued declines in adequacy, although some of the projected need for new transmission may be met by the construction of generating units close to load centers.

To further compound the problem, transmission planning is not keeping pace with the need for such planning. Because transmission owners and ISOs are receiving so many requests for generator interconnections, they are unable to devote the staff resources needed to develop proactive transmission plans. That is, they are focused primarily on preparing the system-impact and facility studies required for these new interconnections. Thus, some transmission plans are little more than compilations of individual generator-interconnection studies.

Because transmission planners have insufficient time and resources, little information is being provided proactively to energy markets on the costs and locations of congestion. Such information could help potential investors in new generation decide where to locate new units. Such information could also help loadserving entities decide what kinds of dynamic pricing and load-reduction programs to offer customers in different locations. More broadly, such information could reduce the need for centralized planning and construction of new transmission facilities.

Because generation and load can serve, in some instances, as viable alternatives to new transmission, transmission plans need to explicitly consider such nontransmission alternatives. Whether the transmission system (i.e., transmission users in general) should pay for these generation and load projects is unclear and hotly contested. At a minimum, transmission planners should provide information (again based on analysis of past and likely future congestion costs) on suitable locations for new generation and load management. In a similar fashion, alternative methods for pricing transmission services (including charges for access, congestion, and losses) would affect transmission uses. These changes in transmission flows would, in turn, affect the need for new facilities. Thus, transmission planning should include assessments of alternative pricing methods to improve efficiency in transmission utilization.

Transmission planning may be too narrowly focused on NERC and regional reliability-planning standards. That is, transmission planning may pay insufficient attention to the benefits new transmission investments might offer competitive energy markets, in particular, broader geographic scope of these markets (which would encourage greater diversity in the fuels used to generate electricity) and a reduction in the opportunities for individual generators to exercise market power. Although some plans consider congestion (either congestion costs or curtailments and denial of service), such considerations are more implicit than explicit. As shown here, congestion costs (both in real time and in forward markets) can provide valuable information on where and what to build.

Advanced technologies offer the hope of better information on and control of transmission flows and voltages. Such improved information and control would permit the system to be operated closer to its thermal limits, thereby expanding transmission capability without increasing its footprint. Thus, new technologies may reduce fights about transmission siting. In addition, these technologies, because they permit control of power flows over individual elements (e.g., DC lines), may make it attractive for private investors to build individual facilities (merchant transmission). Unfortunately, these advanced technologies are still too expensive for widespread application, although some are economic in niche applications.

The separation of generation from transmission and of retail service from transmission poses difficult information problems for transmission planning. Specifically, transmission planners need detailed information on the timing, magnitudes, and locations of new generating units; the developers of these facilities are unwilling to share competitive information until required to do so (e.g., for environmental permits and for transmission- interconnection studies). Planners also need detailed information on the locations and magnitudes of future loads. In a retail-competition world, it is not clear what entities will have the information necessary to produce reliable projections of retail load and whether those entities will be willing to share these projections with transmission planners.

Finally, the economies of scale in transmission investment argue for overbuilding, rather than underbuilding, transmission. It is substantially cheaper per GW-mile to construct a higher-voltage line than a lower-voltage line. The higher-voltage line also requires less land per GW-mile, which should reduce opposition from local landowners and residents. Also, building a larger line now eliminates the need to build another line in several years. This situation can eliminate the need for another potentially bruising and expensive fight over the need for and location of another transmission line. Also, the availability of suitable land on which to build transmission lines can only go down in the future, as population grows and the economy expands. On the other hand, overbuilding can increase financial risks for the transmission owners.

Acknowledgments

We thank John Adams, Kenneth Donohoo, Philip Fedora, Trudy Harper, Michael Henderson, Steven Herling, Douglas Larson, George Loehr, Paul McCoy, Mark MacLeod, Stephen Metague, David Meyer, Wayman Robinett, Jose Rotger, Richard Sedano, Alison Silverstein, Brian Silverstein, Dorothea Stockstill, Perry Stowe, and James Whitehead for their very helpful comments on a draft of this paper.

References

American Electric Power Service Corp. et al. 2000. *Alliance Companies Compliance Filing*, Docket Nos. ER99-3144-003 et al., submitted to the Federal Energy Regulatory Commission, Columbus, OH, September 15.

American Transmission Company. 2001. *Initial Ten Year Transmission System Assessment*, Waukesha, WI, June 1.

Avista Corp. et al. 2000. *Supplemental Compliance Filing and Request for Declaratory Order Pursuant to Order 2000*, Docket No. RT01-35-000, submitted to the Federal Energy Regulatory Commission, Spokane, WA, October 16.

Bonneville Power Administration. 2001. "Transmission Integration Requests Pour In," *Journal*, DOE/BP- 3365, Portland, OR, March.

Detmers, J., A. Perez, and S. Greenleaf. 2001. "Valley-Rainbow Transmission Project," Memorandum to California Independent System Operator Board of Governors, Folsom, CA, page 3, March 23.

Emerson, A. and J.D. Smith. 2001. *Revised Biennial Transmission Assessment 2000–2009*, Arizona Corporation Commission, Phoenix, AZ, July.

ERCOT. 2001. *Report on Existing and Potential Electric System Constraints and Needs within ERCOT*, Electric Reliability Council of Texas, Taylor, TX, October 1.

GridFlorida, LLC. 2000. *Attachment N, Planning Protocol, Supplemental Compliance Filing*, Docket No. RT01-67-000, submitted to the Federal Energy Regulatory Commission, December 15 (as revised on May 29, 2001).

Hirst, E. and B. Kirby. 2001. *Transmission Planning for a Restructuring U.S. Electricity Industry*, Edison Electric Institute, Washington, DC, June.

Howe, J.B. 2001. "End the Gridlock: Why Transmission is Ripe for New Technology," *Public Utilities Fortnightly* 139(2), 38-43, page 40, January 15.

Infrastructure Technical Review Committee. 2001. *Upgrading the Capacity and Reliability of the BPA Transmission System*, August 30.

ISO New England. 2001. *2001 Regional Transmission Expansion Plan (RTEP01)*, Holyoke, MA, October 19.

Liles, J. 2001. "Merchant Transmission: Building a Grid that Wall Street Can Understand," *Public Utilities Fortnightly* 139(17), 24-36, September 15.

Mazur, R.W. 1999. "*MAPP Regional Transmission Planning*," Manitoba Hydro, Winnipeg, Manitoba, November.

Mid-Continent Area Power Pool. 2000. *MAPP Regional Plan 2000 through 2009*, Minneapolis, MN, December 29.

National Grid USA. 2000. *National Grid USA Five Year Statement for the years 2001 to 2005*, Westborough, MA, December.

National Institute of Environmental Health Sciences. 1999. *Health Effects from Exposure to Power-Line Frequency Electric and Magnetic Fields*, No. 99-4493, National Institutes of Health, Research Triangle Park, NC, June.

New England Transmission Owners et al. 2001. *Joint Petition for Declaratory Order to Form the New England Regional Transmission Organization*, Docket No. RT01-86, submitted to the Federal Energy Regulatory Commission, Holyoke, MA, January 16.

North American Electric Reliability Council. 2001. *Transmission Expansion: Issues and Recommendations*, Transmission Adequacy Task Force, Draft Report to the NERC Planning Committee, Princeton, NJ, September 17.

North American Electric Reliability Council. 1997. *NERC Planning Standards*, Princeton, NJ, September.

Pasternack, B. 2001. Personal communications, Transmission Planning Department, American Electric Power, Columbus, OH, March.

PJM Interconnection. 2001a. *Final Draft, 2000 Baseline RTEP Report For the 2002-2007 Period*, Norristown, PA, March 1.

PJM Interconnection. 2001b. *PJM Regional Transmission Expansion Plan (RTEP)*, Version 2.1.00, Norristown, PA, August.

Rotger, J. and F. Felder. 2001. *Connecting Customers: A Case for Merchant Transmission with a Regulatory Backstop in a Restructured Electric Power System*, TransEnergie U.S. Ltd.

Sanford, M., V. Banunarayanan, and K. Wirgau. 2001. *Implications of Capacity Additions in New York on Transmission System Adequacy*, GE Power Systems Energy Consulting, prepared for New York Independent System Operator, Schenectady, NY, March 2.

Seppa, T. 1999. "Improving Asset Utilization of Transmission Lines by Real Time Rating," presented at T&D Committee Meeting, IEEE/PES Summer Power Meeting, Edmonton, Alberta, Canada, July 22.

Southern Company Services. 1995. *Guidelines for Planning Transmission Facility Improvements for the Operating Companies of the Southern Electric System*, Birmingham, AL, March 14.

Texas Public Utility Commission. 2001. *2000 Annual Update on Activities in the ERCOT Wholesale Electricity Market*, Project No. 19616, Austin, TX, March.

U.S. Department of Energy Task Force on Electric System Reliability. 1998. *Maintaining Reliability in a Competitive U.S. Electricity Industry*, Washington, DC, page 126, September 29.

U.S. Federal Energy Regulatory Commission. 1999. *Regional Transmission Organizations*, Order No. 2000, Docket No. RM99-2-000, Washington, DC, December 20.

VanZandt, V. 2001. "Transmission Assessment," Transmission Operations and Planning, Bonneville Power Administration, Portland, OR, March 23.

Western Governors' Association. 2001. *Conceptual Plans for Electricity Transmission in the West*, Denver, CO, August.

Whitehead, J.T. 2001. Personal communication, Transmission Planning Department, Tennessee Valley Authority, Chattanooga, TN, Oct. 17.

Winter, T.M. and K. Fluckiger. 2000. "Policy Issues Regarding Alternatives to Transmission and Recommendations on Tri-Valley RFP," Memorandum to California Independent System Operator Board of Governors, Folsom, CA, page 4, April 18.

Transmission Siting and Permitting

(National Transmission Grid Study)

David H. Meyer
Consultant, Electricity Policy and Regulation
Alexandria, Virginia
Richard Sedano
Regulatory Assistance Project
Montpelier, Vermont

INTRODUCTION

In order to construct new transmission facilities or to significantly upgrade existing facilities in the U.S. electricity system, developers typically need approval from several state and federal agencies. This process has, in recent years, become protracted and difficult. The difficulty is hardly surprising given that transmission facilities are highly visible structures that may span long distances and must somehow fit into physical surroundings that are already in use for other purposes. Incorporating these facilities into the landscape and taking fair account of the wide range of legitimate interests affected by them is challenging.

Nevertheless, many observers and participants in the electricity sector now regard transmission siting and permitting procedures as a major reason why the development of new transmission facilities is not keeping up with the need. Critics say that the siting and permitting process has become unnecessarily cumbersome, delay prone, and subject to breakdown. Some observers argue that current state-based regime for managing siting and permitting is not well adapted to the review of proposed large-scale multistate transmission projects that are or may soon be needed to serve regional bulk power markets, perhaps with little benefit to local electricity consumers. Other officials familiar with state processes agree that regulatory processes can and should be improved, while noting that there is also potential to improve the siting and planning practices of transmission owners or other applicants for proposed new facilities.

Given the vital importance of the transmission network, it is essential to the national interest that transmission siting and permitting procedures work for society in practical terms. That is, these procedures must lead to timely decisions by appropriate agencies about whether proposed facilities would serve the public interest, and to timely approval of routes or sites for facilities that are deemed necessary. This paper examines current siting and permitting practices and ways to improve them. Specifically, the paper:

- Examines existing government and industry practices related to siting and permitting,

- Identifies key or frequent problems with these practices,

- Identifies policy options that should be considered to resolve these problems, and

- Discusses objectively some of the advantages and disadvantages of the options, so they can be considered by federal and state policy makers, corporate officials, and the public.

The policy options discussed fall into three categories:

- Creating new regional institutions to facilitate transmission siting and permitting, either for all new transmission facilities or for large or critically important facilities;

- Improving the current state-based governance regime;

- Making siting-related practices by industry and government agencies more effective, regardless of governance structure.

The remainder of this paper is divided into six sections:

- An assessment of the existing state-based siting regime.

- A discussion of transmission siting from a regional perspective, the reasoning that has led to increased interest in establishing regional institutions for siting new transmission facilities, and the options for designing these institutions.

- Issues related to defining "regional transmission facilities." Some such definition would be useful for determining which new facilities would be subject to the jurisdiction of regional siting institutions.

- Options for improving the existing state-based siting regime.

- Options that could be pursued under any governance structure to improve siting-related practices by government agencies and industry.

- Summary and conclusions.

ASSESSMENT OF CURRENT SITING REGIME

The North American electricity grid is a monumental feat of imagination, planning, and engineering. The grid links generators to cities; cities are linked together and with rural areas; and many electricity suppliers are made accessible to users. Networking delivers a very high standard of reliability at reasonable cost, and the U.S. economy depends heavily upon this high level of reliability. Some government authority approved construction of most of the power lines that make up the grid.

Siting transmission lines is understandably difficult, involving complex engineering, social, and land use considerations. As aggregate electricity usage in an area grows, reliability tends to degrade unless the transmission network is strengthened. There are often many ways to meet a need for grid enhancement, and choosing a good solution is likely to involve tradeoffs among many factors, including cost. Arriving at good solutions often requires long lead times and the development and implementation of a flexible long-term plan for optimizing the transmission grid and related facilities.

Utilities, whether publicly or privately owned, involve a mixture of public and private interests. One of their roles is to bring forward proposals to meet transmission grid needs and implement these proposals if they are approved by government agencies. Consumers rely on government agencies to select the transmission proposals that are most likely to have value well in excess of their cost over the working life of the investment. Because the future is uncertain and reliance on forecasts is unavoidable, the selection process will not always result in the best decisions. However, the goal is that the system used for siting electric transmission lines will produce timely, high-quality decisions in most cases.

Siting electric transmission lines is currently a state responsibility.[1] Each state may address transmission siting in its laws, and most have done so. In a few states, utilities are required only to give notice of intent to build a transmission line; after a specified period, if no challenges are raised, the utility may proceed with acquisition of right of way (if needed) and construction. In most states, however, the utility

---

[1]With the exception of the federal power marketing administrations and the Tennessee Valley Authority, which have their own siting authorities.

must demonstrate to a siting authority that the proposed facility is needed, and the siting authority must confirm that construction of the facility will serve the public interest.

Most power lines are proposed in states that have formal siting authorities. Some transmission proposals are withdrawn after supporting evidence is assessed during the siting process. A few proposals make it all the way to a decision by the siting authority and are then rejected. Rejections represent failures of analysis and communication somewhere in the planning and siting processes, and they are costly to all parties, including the public. The objective of the subsections below is to examine the existing siting process and analyze some of its successes and failures.

Description of the Transmission Siting Process

A utility typically files a siting proposal when it feels that there is justifiable need for additional transmission capacity and that the proposed solution is robust. In most states, the proposal goes to a siting authority, most often the regulatory utility commission. A significant number of states have a separate siting authority that may include officials from other affected state agencies.

Usually, the process is a "contested case," which means that the decision will be based on evidence presented by the applicant and other parties. Parties ("intervenors") may intervene in the case either by right (e.g., the state public advocate) or by permission if they demonstrate to the siting authority that a distinct interest is at stake that is not otherwise sufficiently represented. The utility decides when the process starts and controls most of the relevant information. Sometimes, intervenors fill gaps in the information provided by the utility.

In some states, a specific amount of time is allotted for reviewing a transmission siting proposal. The time limit may be reached or even exceeded in complex cases or cases that involve much procedural maneuvering; this may trigger a rejection of the proposal by the siting authority on procedural grounds. Other states have no specific time limit. Still other states, in order to reduce utility incentives to hold back details about a proposal, allow a time limit to be activated only after a finding by the siting authority that the application is complete. In some states, the process focuses on the proposal under consideration rather than on how best to address a grid need.

In these cases, a rejection may not be accompanied by guidance about how to address better the need that the original proposal was intended to meet. The prospective lack of such guidance and the desire to avoid rejection may motivate some parties to work during the case to improve the project after filing, based on evidence and arguments during discovery and hearings.[2]

Electricity consumers pay for transmission facilities through their electricity bills.[3] Consumers depend on regulators to allow the incorporation into electricity rates only the costs related to transmission facilities required to serve their area's long-term needs. Transmission costs represent approximately 10% of the nation's total electric bill.[4] An environmental assessment is often required for a transmission proposal. Environmental issues of interest include:

- Concern about opening new areas to development—for example, roads may be needed for access to maintain lines, and development may follow roads;

---

[2]An iterative process has its merits but exposes intervenors to the risk of having to evaluate an essentially new proposal in the midst of the process. The siting authority must "manage the clock" to ensure that everyone is treated fairly.

[3]Merchant transmission costs find their way into retail prices though by a different path than regulated transmission rates.

[4]DOE. 2002. National Transmission Grid Study. U.S. Department of Energy.

- Potential disruption of habitat by reducing the size of continuous undeveloped spaces;

- Potential impacts on endangered species; and

- Visible impacts that may create aesthetic concerns, especially in scenic areas.

In most states, the utility must apply for and obtain a "certificate of public need" (the name of this document varies from state to state) for a transmission facility. This certificate is extremely important; it indicates that the designated government authorities have reviewed the proposed project, evaluated the tradeoffs involved, and concluded that, overall, the project is in the public interest even though some legitimate private or public interests may be adversely affected. The formal criteria for determining "need"[5] vary. Some commonly used criteria are

- Someone is willing to invest in the project (in other words, the project is perceived to have significant marketable value).

- The project is needed to maintain the reliability of the bulk power supply system.[6]

- The project is needed for regional electricity commerce.[7]

---

[5]There is no practice or mechanism for determining regional or interstate need. The Electric Reliability Council of Texas (ERCOT) performs this function in a way that some expect will become typical for Regional Transmission Organizations (RTOs)—providing unbiased and competent information to clarify and focus the work of individual utilities on addressing validated needs.

[6]Demonstrating this particular need requires competence in either deterministic or probabilistic transmission planning models as explained briefly in Section 6. Using just one approach leaves the applicant vulnerable to challenge.

[7]For some states, serving regional commerce is a vital purpose of the grid. For others, it is secondary to maintaining reliability.

- The project is needed to interconnect an approved generator to the grid.[8]

In many states, decision makers must consider alternatives to the primary proposal. Some states have specific instructions concerning alternatives that the utility must present. Siting authorities are typically interested in route and non-transmission alternatives when these are relevant.[9]

Substitutability of
Transmission and Nontransmission Resources

There are many substitutable ways to meet customer needs for delivery of energy. Here are two examples:

1. Consider a community that has experienced significant customer demand growth and has been relying on generation located outside the area but delivered to customers by wires that are beginning to reach capacity limits. In this case, the capacity of the lines could be increased, or generation could be added within the community to reduce the need for imports. Alternatively, customers could reduce their demand on the grid, either by using energy more efficiently or by making their own electricity. Deploying several approaches may avoid overreliance on any one. Examples of these alternative approaches are being deployed now in New York City for the explicit purpose of improving electricity system reliability.

---

[8]One commenter at the public workshops organized for the National Transmission Grid Study (NTGS) by the U.S. Department of Energy suggested that inadequate attention is being given to transmission needs associated with bringing some new generation on line. If this is true, a need buildup may be accumulating that could result in belated justification for new power lines in some areas.

[9]Texas requires submittal of alternative route options as well as analysis of the usefulness of demand-side management and distributed generation in lieu of new lines.

2.   Consider a market in which a transmission constraint leads to en-
     ergy clearing prices that differ by two cents between the two sides
     of the bottleneck in many hours. Possible solutions include adding
     transfer capacity to allow the low-cost resources on one side of the
     constraint to flow freely to the other side. Or it might be possible
     to add lower-cost resources in the region where energy prices are
     higher. A third alternative would be to reduce demand in the region
     where power is more expensive if a reduction would mean avoid-
     ing use of the most expensive generation resources. An example of
     adding resources in the region where power is more expensive
     appears to be unfolding in Pennsylvania where differences between
     eastern and western prices are moderating because natural-gas-
     fired generation has been added in the east.

In both examples, structural improvements, such as more functional
markets and better pricing regimes, are contributing to the resolution of
problems that might once have been solved by transmission facilities
alone. These alternatives should be considered during an investment
planning cycle prior to and again during permitting so that the public
can see and appreciate the decision-making process.

The cost to prepare a transmission proposal and support
it through the siting process is significant and can vary depending on the complexity of the project and degree of public concern.[10] Regulated companies expect that federal and
state supervised rates will recover the cost of the project plus
a reasonable return. Merchant transmission companies rely

---

[10]In September 1995, the Florida Public Service Commission (PSC) voted to allow
Florida Power to recover $23 million in costs spent on a proposed 500-kV line
that was never built. The line was approved by the PSC in 1984 for reliability.
However, continued local opposition led to protracted and costly litigation.
Florida Power eventually developed an alternative plan involving more intensive
monitoring of the status of key transmission lines in the area, interruption of
service to a limited number of customers in emergency situations if necessary,
and reactivation of a 115-kV line that had earlier been retired from service. (Electric Utility Week, 1995.)

on a business plan that forecasts sufficient revenues from the sale of transmission services to cover their costs and provide an acceptable return on invested capital. Their charges are also eventually reflected in retail electric prices.

A crucial and volatile factor in the transmission siting process is the public trust. It is extremely important that the managers of the process and other major parties act in specific cases so as to gain and keep the public's confidence that the siting process will generally lead to sound outcomes that serve the many public interests at issue.

Due Process in Transmission Siting

Due process is an important element in the American judicial system, including the transmission siting process. By means of due process rules, the regulatory agency that manages the process balances the interests of many parties, including potential intervenors who need a sufficient opportunity to review and critique the particulars of a proposed transmission project, the utility that is charged with providing reliable service at just and reasonable rates, and consumers.

The first element of due process is notice. Parties who may be affected by a project have a legal right to hear about it sufficiently in advance to make a reasoned response if they choose. When a project affects many communities, notice must be provided so that all communities are informed.[11]

A complete filing is also a necessary element of due process. Potential intervenors need full information about the project, presented in non-technical terms. Information provided by utilities may be incomplete. Regardless of the

---

[11]This concern is spoofed in The Hitchhiker's Guide to the Galaxy, by Douglas Adams, in which notice to demolish the Earth was posted at Alpha Centauri.

history or regulatory time limits on the case, filing of incomplete applications or withholding of relevant information puts the proposed project at risk, and may create mistrust, conflict, delay, and/or result in outright rejection of the proposal. Another key element of due process is the determination of which parties are allowed to participate. The state is usually represented, and any relevant point of view not adequately represented by others is generally allowed. Those designated as "parties" to the case receive all information submitted to the siting authority by any other party and have the opportunity to ask and be asked discovery questions and to put on and cross-examine witnesses. Typically, parties pay their own costs. Low-budget participation is possible, but expert advice is expensive, which limits the participation of some intervenors.[12]

"Discovery" is the process of insuring that all relevant facts are available to all parties before hearings. Because utilities possess most of the information relevant to transmission proposals, they usually have a greater discovery burden—that is, they must distribute to others all relevant information. In some cases, however, other parties present competing alternatives and thus become subject to major discovery burdens.

Conflicts may result if some information essential to understanding the need for or the design of a project is declared to be confidential to protect allegedly proprietary details. A simple solution is an agreement that allows all parties to see the information but requires that they use it only for the purpose of the case. Even with such an agreement, disputes may persist since the information may be

---

[12]Some of the most tenacious non-government intervenors have wealthy benefactors or pro bono advocates. In rare cases, states provide funding, usually assessed from the applicant, for intervenors.

important to enable the public to understand the need for the project, and there is no practical way to include the public in a protective agreement. In addition, there may be lingering disagreement on how proprietary the information is in the first place. In many jurisdictions, applicants face no formal penalty if they withhold information as a strategy to divert attention or delay review of the proposal; however, an applicant who withholds information risks losing the trust and goodwill of regulators and the public.

Siting authorities usually allow public comment, and many are required by law to do so. Some states require that comments be solicited in person in each affected county. For a long transmission line, many counties could be affected. Public comments are not usually used as evidence because statements are not cross-examined; however, these comments may influence the atmosphere in which the decision makers deliberate.

Technical hearings are the forum through which the siting authority collects evidence. These hearings are sometimes held before staff or hearing examiners or directly before the siting authority. Parties to these hearings can produce witnesses, and all parties can cross-examine all witnesses.

It is important that all parties understand in advance the standards for approval of a transmission proposal. These standards should be provided by the siting authority with citations of appropriate statutes, regulations, and precedents. Sometimes an issue emerges for which there is no precedent, and parties may want to know at an early stage in the case how the authority will evaluate this issue. After the siting authority issues its findings and orders, there is usually an opportunity to appeal. State courts vary in their ability to process such appeals quickly.

Key Difficulties in the
Current Transmission Siting Process

Why don't utility proposals for new transmission facilities get routinely approved within a "reasonable" time period? In fact, most smaller projects or upgrades of existing facilities are approved, often in less than a year. Notice and hearing requirements take up the bulk of the time in such cases. However, some proposals do not go smoothly, as discussed in the rest of this subsection.

Significant difficulties arise when a proposal is perceived by key parties to be inconsistent with important public interests. These interests may include costs as well as impacts on electric rates, the environment, property rights, protected federal land, or other sensitive land. Often, critical disagreements are about how certain tradeoffs should be evaluated and resolved. Sometimes, a conflict is the result of a party's conscious decision to be uncompromising for reasons of principle or strategy. Disputes may arise about whether certain questions have been sufficiently answered or whether parties will have access to certain information and on what terms.[13]

Major delays occur if the siting authority finds that an applicant failed to examine and present relevant alternatives, a task that entails significant effort. If more than one state is involved, the states may disagree over the proposed distribution of the societal benefits and costs associated with the line.

A bias is introduced in the weighing of alternatives if different approval venues, processes, or compensation meth-

---

[13]There are many examples. In Illinois, a transmission project was approved only after the utility produced information requested by the commission staff; the staff had recommended that the project be denied because the information offered at the outset was inadequate. Illinois Commerce Commission Docket 92-0121 (P.R. Buxton, Personal communication).

ods are used for different options. For example, if the siting authority is not the regulatory commission, the authority may not have sufficient experience in demand-side measures to determine whether they may be superior to a power line as a means of meeting a system need. Introducing competition to the wholesale generation market has added another dimension of difficulty. Investments in generation, transmission, and demand-side measures come in regulated and competitive forms and pass through different channels for approval, so there is no single standard for comparing them, and there may be no formal opportunity for a side-by-side evaluation.

Two Instructive Transmission Siting Cases

American Electric Power's 765-kV project between
West Virginia and Virginia

The painfully long, complex, controversial, and costly review of an American Electric Power (AEP) transmission project in West Virginia and Virginia is often cited as a definitive example of a dysfunctional transmission siting process. The major parties are the applicant, two states, and three federal land management agencies. After ten years of review, this project is still at least a year from final approval.

AEP first proposed the 765-kV project in 1991 to Virginia, West Virginia, the U.S. Forest Service, the National Park Service, and the U.S. Army Corps of Engineers. As initially proposed, the project's primary purpose was to maintain reliability in southern West Virginia and southwestern Virginia, and a secondary purpose was to reduce the risks of a cascading outage that could affect many states in the eastern United States. The project would have involved construction of a new line about 113 miles long from an AEP substation in Wyoming County, West Virginia, to an AEP substation near Cloverdale, Virginia. Possible impacts on populated areas made the project controversial in both states, and both states held very extensive local hearings. In addition, the Forest Service issued a draft environmental impact statement in 1996 in which it recommended that the line not be constructed as proposed because it would cross sensitive areas of the Jefferson National Forest, the Appalachian Trail, and the New River.

In October 1997, AEP proposed an alternative route to the regulatory commissions in the two states. This route was about 17 miles longer than the earlier route, and the most important change was that it would go south from the Wyoming area of West Virginia before turning east, enabling the line to cross the New River in a less sensitive area. Several other changes were made to put the line behind ridges and to cross rivers and important natural areas at locations with lesser impacts. In June 1998, the West Virginia Public Service Commission approved its 32-mile portion of the line.

In September 1998, however, AEP agreed to a request from the staff of the Virginia Corporation Commission that the utility conduct a detailed study of an alternative route that would follow much the same path as before in West Virginia but would terminate in Virginia at an AEP substation near Jacksons Ferry. The Virginia Commission also engaged a consulting firm to prepare an independent evaluation of the route to Jacksons Ferry. After completing its review, AEP agreed that the Jacksons Ferry route was acceptable although it would not allow as much margin for future load growth as the route to Cloverdale.

In May 2001, the Virginia Corporation Commission approved the Jacksons Ferry route, chiefly because it would have fewer adverse environmental and social impacts than the route to Cloverdale. The West Virginia Public Service Commission must now review the route ending at Jacksons Ferry, even though the West Virginia portion of the route remains essentially unchanged from that which the commission approved in June 1998. In addition, the new route would cross about 11 miles of national forest in an area not studied in the Forest Service's 1996 draft environmental impact statement, so the Forest Service must do a supplementary analysis and decide whether to grant a permit for construction of the line.

The siting process for this project might have been accelerated if there had been:

- Greater coordination and cooperation among the five reviewing agencies (West Virginia, Virginia, and the three federal agencies). A significant source of delay in the earlier stages of the process was that each state commission tended to favor a route that would reduce adverse environmental and social impacts within its own state without regard for the possibility of adverse impacts in the other state.

- Presentation by AEP of a wider range of alternatives at an early stage in the process.

- Better communication between the Forest Service and the applicant. The Forest Service and the applicant could have focused earlier on the acceptability of several alternative routes across national forest lands.

- More emphasis on the "regional picture" through involvement of a regional siting institution. Because a major purpose of the line is to reduce the risk of a cascading multistate outage, this project has regional significance. The regulatory process, however, has involved only two states, and their proceedings have focused primarily on intrastate concerns.

The Cross Sound Connector

Another project, the Cross Sound Connector, illustrates the problems of focusing on a single route and also shows some additional difficulties typical of interstate projects. TransEnergie US, Ltd proposed the project in the summer of 2000. It would connect the Long Island Power Authority's Shoreham substation with a United Illuminating substation in New Haven, Connecticut, by means of a buried 26-mile undersea cable. The project has two principal purposes: to improve reliability on Long Island and in Connecticut, and to enable Long Island to import generation from New England. The project obtained required approval from New York officials but was rejected in April 2001 by the Connecticut Siting Council, less than a year after it was proposed.

Two reasons were cited for the rejection. The primary reason was risk to valuable shellfish beds in Long Island Sound near the Connecticut end of the project. A secondary concern was that the allocation of benefits from the project between New York and Connecticut was not equitable in comparison to the burdens involved. In August 2001, TransEnergie reproposed the project with a new route that would avoid the shellfish beds at some additional cost. Because the first proposal was rejected without prejudice, the revised proposal was filed as a new application in Connecticut, and went through the full review process. The Connecticut Siting council approved the project on January 3, 2002. However, some critics of the project announced their intention to challenge the Council's decision in court.

This case highlights that before filing a formal proposal, an applicant should probe thoroughly for sensitive issues that may be raised by its proposal and the likely impacts of alternative routes. The case also demonstrates the need for states involved in the review of interstate projects to coordinate their reviews and agree on findings regarding the allocation of costs and benefits. (These topics are discussed below in the Regional Perspective section.)

Successes in Siting Transmission

Most transmission siting proposals eventually receive certificates of need. With sustained effort, utilities, state regulators, public advocates, communities, and intervenors usually find answers to problems. A successful review process for a large interstate transmission project is described below.

A recent four-state transmission siting success story

In September 1998, New Century Energies (a company formed by the merger of Southwestern Public Service and Public Service of Colorado and subsequently merged into Xcel Energy) affirmed its intent to build a 300-mile, 345-kV line that would connect a Southwestern substation near Amarillo, Texas, with a substation near Lamar, Colorado, that is partially owned by Public Service of Colorado. From Amarillo, the line would cross the Oklahoma panhandle, continue north to Holcomb station near Garden City, Kansas, and then west to Lamar. The terminus at Lamar was to be a 210-MW high voltage DC interchange facility that would permit asynchronous flows between the eastern and western U.S. grids. The purposes of the project were to improve reliability and stabilize power flows in the region and to facilitate electricity trade. To address potential market power concerns associated with the company merger, Texas regulators required New Century Energies to pursue this project. In July 2001, Xcel Energy obtained the consent from the last of the four states when the Colorado Public Utilities Commission approved the project.

The interest of Texas regulators in this project only partly explains the project's success. Other reasons were the applicant's proactive anticipation of and responsiveness to landowner and community concerns, and the awareness by Kansas regulatory officials of the regional implications of the project and the potentially reciprocal responsibilities of a state faced with a project of principal benefit to neighboring states.

Critical Elements of Success and Conditions that May Lead to Conflict

A review of many siting proposals reveals some indicators of probable success as well as conditions that increase the risk of conflict.

Success indicators

- Link to a generation project—A transmission project that interconnects a needed generation project to the grid is less likely than other types of projects to encounter heavy opposition. The transmission component may be seen as incidental to the generation project.

- Early planning—If interested parties are informed ahead of time that a power line may be needed and will probably be proposed, the project has a greater likelihood of success. In some cases the proposal that is ultimately put before siting authorities differs from that which had initially been presented to the public for review, indicating that the public review was of value.

- Open planning—A planning process is considered "open" or "transparent" when it solicits the views of interested parties regarding ways to address a specific transmission need. Parties other than utilities are more likely to feel that such a process has respected their interests; it also gives the utility the opportunity to make changes to a plan before committing to it as a formal proposal.

- Regional planning—The major benefits from interstate transmission projects are often unevenly distributed. When out-of-state benefits can be recognized in a state's siting process, effective presentation of these benefits is an important indicator of success. Special arrangements may be needed to ensure that a project will provide net benefits to all affected states.

- Demonstrable need—A project appears more compelling as its value to consumers is more evident. The need to maintain reliability is widely accepted although demonstrating that a specific project is needed to strengthen the system can be difficult. The need to interconnect a permitted generator to the grid is usually obvious. In some states, there is debate about the "need" for projects that primarily facilitate electricity trade.

- Economic benefits—If regional energy transfers are clearly in the public interest, a proposed project that enables such transfers will likely be received positively. In some jurisdictions, applying this rationale to power lines is relatively new and results from the increasing importance of wholesale electricity trade. There is debate about whether a proposed transmission line that primarily facilitates electricity trade and reduces electricity costs for some consumers is "needed" (see Regional Perspective section for further discussion). In some jurisdictions, "need" is interpreted narrowly as referring only to reliability.

- Alternatives, presented objectively—Presenting a broad

range of relevant alternatives is important. Some states require that alternatives accompany the primary siting proposal, and intervenors and public advocates may develop them if the utility does not. Regardless of Transmission Siting and Permitting **E-11 E-12** National Transmission Grid Study regulatory requirements, an objective presentation of alternatives advances the credibility of the applicant and the primary proposal.

- Open lands—If there are few objections to the transmission line route on the basis of natural resource concerns, the odds of a project's success improve, particularly as open land area shrinks in many states with the growth of cities. (Restrictions on the use of much government land limits its value for transmission siting.)

Characteristics of transmission siting proposals/processes
that may lead to conflict

- Disregard for directives in law or siting authority pronouncements—Probably the worst thing that a transmission project applicant can do is disregard clear instructions from the siting authority or statute. Although this may seem unlikely, it happens more often than might be expected.

- Differing assumptions about land use—State officials may view proposed land use tradeoffs in ways that differ from utility expectations. An open utility transmission planning process can reveal potential misunderstandings of this kind before they disrupt or derail a mature proposal.

- Potential for disagreements with federal land managers—Difficulties sometimes arise when the interests of one or more federal land management agencies are affected by a proposal. Land managers may not regard accommodating transmission line proposals as a high priority. Different federal land managers within a region may not coordinate well with the state siting process or with each other, even within a single federal department. Federal land managers sometimes decide not to commit resources to participate in the planning of a transmission project (or ignore the process, which has the same result), choosing to participate only after the process is well under way, compromises have been made by others, and the range of options under consideration has been narrowed. Some projects affect the interests of several federal agencies, and some parties cite insufficient coordination among them in reviewing such projects as a problem. (Note: There are also cases in which these managers have cooperated well with each other and with state siting officials.)

Business Uncertainties and the Current Siting Process

The business aspects of the current transmission siting process merit attention. The ongoing restructuring of the U.S. electricity industry poses many uncertainties for the transmission component of the industry. Some companies do not know whether they will remain in the transmission business, and those that intend to stay in the business are unsure what rules will determine the profitability of new transmission investments. There is also uncertainty about how market participants will gain access to transmission facilities, and receive allocations of scarce transmission capacity. The outcomes of these federal legislative and regulatory debates will create winners and losers, and the debates are a consuming preoccupation for participants at all levels of the electric industry.

Some parties believe that many meritorious transmission projects never make it out of the utility board room and into the permitting process. It is unclear whether this is because of uncertainty about whether revenues will cover the cost of the facility, skewed incentives resulting from unsound transmission pricing, fear of the siting process, faulty project development, concern for predatory effects on profits from other utility investments (i.e., generation), other reasons such as local politics, or some combination of influences. The cost of the siting process weighed against the odds of success is understandably important. It is equally important, however, to remember that, in every part of the United States, there is an entity obliged to deliver electricity reliably and at a just and reasonable rate. This obligation does not account for business risk though first principles of regulation call for utilities to be treated fairly by being given the opportunity to collect adequate revenues for their service. These entities must continue to try to build the facilities they believe are needed.

THE REGIONAL PERSPECTIVE

Since the 1970s, electricity providers have increasingly used the nation's transmission networks for electricity trade as well as for the traditional purpose of ensuring the reliability of bulk power supplies. During the past decade, electricity trade has increased very sharply, to the point that congestion is now frequent in many locations and economically desirable trades must often be foregone to avoid loading the transmission lines beyond prudent limits.[14] In

---

[14]New tools for managing the grid may enable operators to maintain reliability standards while reserving less transmission capacity for contingency flows. This will relieve constraints in some areas at some times.

addition, as the aggregate economic value of the trade enabled by the grids increases, the trade function becomes increasingly important, and the two functions of maintaining reliability and enabling trade tend to converge. From the perspectives of transmission planning and operations, the overall goal is now to facilitate trade while maintaining reliability.

Although many states do not now take electricity trade into account when issuing permits for new transmission capacity, this may change. In general, all levels of government (federal, state, and local) have long since adopted the policy premise that additional commerce enhances productivity and serves the public interest, assuming that the prices for the goods and services involved accurately reflect real costs. Attention to the externalities or dislocations that could result from trade often leads to requirements for mitigation, and in some cases to outright rejection of proposed additions to an area's infrastructure. Further, if insufficient attention is given to adverse side effects of increased trade, the probability of misallocated or excessive investment goes up markedly. For example, excessive transmission investment could be underutilized because of electrical stability concerns, or excessive investment in local generation could cause generation to be "locked into" a region. A thoughtful assessment of alternatives, as discussed in the section "Improving Agency and Industry Practices," on page E-31, helps to ensure the broad vision necessary to consider all aspects of additional electricity commerce in transmission planning and siting processes.

In any case, given that the policy of favoring increased trade has won broad acceptance, it seems likely that states will increasingly acknowledge the contribution of electricity commerce to the need for new transmission capacity. Given the long-term and forward-looking nature of transmission planning, planners should take into account likely future trade requirements even if some jurisdictions in their area do not now recognize trade as contributing to need. Some analysts note that the reliability benefits of transmission additions are typically distributed very broadly, and the costs of such additions are usually recovered from all consumers across a wide area; by contrast, the economic benefits of increased commerce may be distributed much less evenly. This means that different methods of cost allocation and recovery may be appropriate, to the extent that a project is needed to support electricity commerce.

Finally, it is apparent that, in general, the public will benefit if the geographic markets across which bulk power trade occurs and reliability is managed are large. This is because large markets tend to be more diversified than small markets, and greater diversity translates into both lower market-clearing prices and lower-cost provision of reliability. (See Issue Papers *Transmission Planning and the Need for New Capacity* by E. Hirst and B. Kirby and *Alternative Business Models for Transmission Investment and Operation* by S. Oren, G. Gross, and F. Alvarado for additional analysis.)

The importance of thinking about bulk power markets in terms of large multistate regions is widely recognized (Fox-Penner, 2001; Bailey and Eaton, 2001; Costello, 2001; O'Donnell, 2000; Stavros, 2000). However, efficient regional markets will not evolve through market transactions alone. Sustained, conscious efforts are needed to develop regional institutions that will support the functioning of such markets. In its Order No. 2000, the Federal Energy Regulatory Commission (FERC) stressed the benefits of large markets; in that and subsequent orders, FERC has emphasized the importance of forming large Regional Transmission Organizations (RTOs). RTOs may be for-profit Independent Transmission Companies (ITCs, also called TRANSCOs), nonprofit operators of transmission facilities owned by others (Independent System Operators, or ISOs), or some hybrid of the two. (For extended analysis of RTOs, see Issue Paper *Alternative Business Models for Transmission Investment and Operation* by S. Oren, G. Gross, and F. Alvarado.) In Order No. 2000, FERC sees large RTOs as essential mechanisms for achieving several transmission objectives that are very important to the public interest, including:

- Provision of nondiscriminatory transmission service to all buyers and sellers in the market area,

- Economically efficient provision of ancillary services,

- Economically efficient assurance of reliability, and

- Regional transmission planning.

Many observers now believe that transmission grids can be planned, built, maintained, and operated most efficiently from a regional perspective. In addition, many are also concerned that the existing state-based regime for siting and permitting new transmission projects may not be well suited to assessing proposals of regional importance. Some of the issues raised are

- The societal costs and benefits of a regionally important transmission project are seldom distributed evenly across the area affected. Benefits tend to be distributed broadly in the form of lower electricity prices, higher reliability, and larger sales volumes for lower-cost electricity producers. By contrast, many costs are distributed narrowly along the route of the proposed line where aesthetic vistas, real estate values, and land use patterns are likely to be negatively affected. In addition, the consumers who pay for the line through their electric bills may or may not be the same group of consumers who benefit from increased reliability and access to lower-cost generation.

A Case of a Failure to Communicate

The siting proceedings described below for a generation and transmission project that had regional impact demonstrate how communication can go wrong among two states and a federal regulator, and how ignoring a project's regional dimensions in the early stages can cause difficulties later.

In 1989, FERC granted the city of Jackson, Ohio, a license to construct a hydro generation project on the Ohio River. AMP-Ohio, a wholesale power provider to 77 Ohio municipal utilities, joined the project as a co-developer and helped finance the project. A decision was made to site the project at Belleville, West Virginia, to take advantage of a West Virginia law that exempted municipal hydro projects from state tax. However, because the economic benefits of the Jackson project would go mostly to retail consumers served by AMP-Ohio's utility customers, controversy arose in West Virginia where it appeared that citizens would suffer environmental impacts but few economic benefits. Accordingly, the West Virginia Senate passed a bill in 1994 removing the tax exemption for the project and threatening its economic viability. Although the governor of West Virginia vetoed the bill, saying that it was unfair to treat out-of-state municipalities differently from those of West Virginia, an agreement was reached before the veto that the project sponsors would make payments to West Virginia in lieu of taxes and that the transmission line linking the hydro plant to the grid would be located entirely wholly in Ohio even though that would approximately double its length.

In 1996, Ohio regulators approved the transmission line, but Ohio Public Utilities Commission (PUC) chair Craig Glazer filed a "concurring opinion" strongly criticizing the review process and its outcome. Glazer complained that Ohio was not consulted "in a meaningful way" when AMP-Ohio negotiated its deal with the governor of West Virginia: "It is indeed disingenuous for AMP-Ohio to reach an agreement with the West

Virginia governor to site the line in Ohio and only then come to Ohio and argue that any routes in West Virginia are not feasible and should not be looked at in the siting process" (Electric Utility Week, 1996). Glazer argued that analyses showed "far more environmentally benign and cost-effective routes through West Virginia for this line." He criticized FERC, which had approved the proposed hydro facility, saying that Ohio staff had attempted to establish a joint siting and information sharing process that "fell on deaf ears at the FERC staff level." He continued, "Given FERC's utter lack of interest in such a cooperative effort, [Ohio's] staff did not pursue more formal requests" for cooperation. He added, "This is a case study on how applicants, neighboring states, and an intervening federal agency should not act" (ibid.).

Although there was a good faith effort to resolve the benefit allocation issue between Ohio and West Virginia in this case, the transmission line was not considered at that time by AMP-Ohio, so the company was vulnerable later to the assertion that it had struck an unscrupulous bargain with the governor of West Virginia. PUC Chairperson Glazer noted that some of these difficulties might have been foreseen at the time of the original hydro licensing decision and could have been resolved in advance. Perhaps due in part to this case, Ohio recently adopted a streamlined, time-limited siting process that explicitly provides for cooperation with other states and agencies on siting matters.

- At least one state is legally prohibited from considering out-of-state benefits associated with projects under review (Mississippi State Code 77-3-14). This constraint could lead to rejection of regionally beneficial projects if the intrastate benefits do not appear to exceed the intrastate costs.

- Even if a state is not legally prohibited from taking out-of-state benefits into account, it may still not give these benefits full weight when assessing a project.

- Existing siting processes vary significantly from state to state. Approval may be required from federal agencies charged with the management of public lands; this is particularly frequent in the West. Permits for crossing the lands of Native American tribes may also be needed. Thus, the review process for a major interstate project is almost certain to be complex. Institutional mechanisms are needed to improve communication and coordination among the various agencies that must approve a project and to help develop common procedures and requirements to serve the needs of as many reviewing agencies as possible.

The concerns noted above regarding the adequacy of the existing state-based process for reviewing major interstate transmission proposals have led some observers to conclude that strong regional authorities are needed to organize reviews and decide about siting and permitting of projects that would have regional impacts. For example, see DOE (1998), Recommendation #25: "Explore formation of regional regulatory authorities (RRAs) to provide an institutional focus on interstate transmission enhancement needs, the avoidance of increased regulatory burdens and the replacement of multiple siting and other authorities with single regional siting authorities that are not subject to any state veto." Note: This recommendation was not supported unanimously.

The principal counterargument expressed by organizations representing state and local government agencies is that as yet there is no compelling evidence that such far-reaching changes are needed. In September, 2001, nine state and local governmental organizations delivered a joint letter to Senator Jeff Bingaman, chairman of the Senate Energy and Natural Resources Committee, objecting to Bingaman's draft legislation that would give the FERC a backstop role and eminent domain authority with respect to siting new transmission facilities. The nine organizations were the National

Governors Association (NGA), the National Conference of State Legislatures, the National Association of Regulatory Utility Commissioners, the Council of State Governments, the National Association of Counties, the National Association of Towns and Townships, the National Association of State Energy Officials, the National Association of State Utility Consumer Advocates, and the Association of State Energy Research and Technology Transfer Institutions (*Electric Utility Week,* 2001).

An examination of recent or current major transmission projects does not yield conclusive answers about whether strong new regional siting institutions are needed (as opposed to improvements to the existing state-based regime). At a minimum, however, the record confirms that new mechanisms and practices are needed to foster greater coordination, cooperation, and timeliness among states, federal agencies, and tribes that must review proposed major interstate transmission projects. Pertinent issues and policy options are discussed in the sections below.

Some Generic Considerations
Regarding the Regional Approach

Before discussing various possible formats for the design of regional siting institutions, it will be helpful to address several background topics that pertain generically to the regional approach.

Relationship between generation siting and transmission siting

Generation and transmission siting are inextricably related. The placement of new generation in relation to load centers and transmission bottlenecks can increase or decrease the need for new transmission facilities. Regional or state planning and siting officials must take these effects into account.

In some areas of the country where natural gas is readily available at low cost (e.g., the Gulf Coast), generation providers have filed applications for transmission interconnections for new generation well in excess of projected load growth in the surrounding area.[15] This generation would serve more distant markets, and additional transmission capacity would probably be needed to enable the generators to reach those markets. However, some parties assert that natural gas pipelines may be generally cheaper and less environmentally intrusive than electric transmission lines, and most analysts agree that new generation capacity should be built as close as practicable to the load centers it serves.

Accordingly, when a new "long line" transmission facility is proposed, opponents may argue that the facility is not needed because new generation could be built near the load center. This would probably raise an evidentiary question (i.e., one requiring formal examination) that would have to be addressed before the question of the need for the transmission facility could be resolved. Further, load centers tend to be heavily urbanized areas; they may have air quality problems; and they may lack the water supplies needed for new generation. Without a thorough assessment of these issues, decision-makers would find it difficult to answer the question of the feasibility in economic and other terms of building a sufficient quantity of new generation near the load center. The need to consider other alternatives to new transmission capacity (e.g. distributed generation)[16] would broaden the analytic requirements of the process even further.

---

[15]See, for example, comments presented by a Southern Company representative at DOE's workshop in Atlanta, September 26, 2001.

[16]A further difficulty is that it takes time, once a need is identified, to combine the many possible resources into a sound mitigating strategy.

This complex of issues (the merits of local generation and other local alternatives versus distant generation plus transmission) has two significant implications:

(1) It increases the prospects for disagreement between or among states concerning the need for new transmission capacity and suggests that states should be cautious about approving new generation capacity without inquiring whether such capacity may lead to transmission congestion and the need for new transmission capacity in neighboring states. The availability of new technologies for distributed generation and other technological substitutes for new transmission will add fuel to this debate. At a minimum, generation and transmission siting decisions increasingly require extensive communication and coordination among states across a region.

(2) It increases the need for open regional transmission planning processes that will indicate to all affected parties where and when new transmission capacity will be needed, taking into account the siting of generation and the economic cost and feasibility of alternatives to conventional transmission facilities.

Promoting common processes among reviewing agencies within a region

A regional institution could foster the development of common processes that all reviewing agencies in the region—states, Native American tribes, and federal agencies—could use to review transmission projects. The regional body could facilitate development of common application requirements and timelines, joint interagency hearings, agreements on the types of alternatives to be considered, and a single record of decision for the project (see *Conceptual Plans for*

Electricity Transmission in the West, 2001). These actions could be accomplished with comparatively little infringement on the authority of the reviewing agencies.

Improving coordination of the overall process in a region

Shortly after an application for siting of an interstate or regionally significant transmission project has been filed with one or more reviewing agencies, it would be beneficial to have a joint meeting involving the applicant and all affected reviewing agencies, including federal agencies and Native American tribes, to identify possible points of difficulty or disagreement and begin exploring possible solutions. Although this meeting could be coordinated informally under the existing state-based review regime, a centralized regional organization could give the effort focused and pragmatic leadership without infringing on the authority of the reviewing agencies.

Two current and controversial transmission siting cases involving Minnesota and Wisconsin[17] provide support for the view that the siting process for interstate projects could be aided significantly if a cooperative regional body were available to assist in coordinating the process, and if regional transmission plans were available to guide state agencies in considering questions related to the need for new transmission facilities. In both of these cases, the applicants contended that the lines were needed primarily to maintain reliability in Wisconsin. The need issue became a matter of debate in both cases, and resolu-

---

[17]These are the 38-mile line from Chisago, Minnesota, to Apple River, Wisconsin, and the roughly 230-mile line from Duluth, Minnesota, to Wausau, Wisconsin. The latter has been approved by the state siting authorities, but is the subject of an appeal in Minnesota. The former was withdrawn and is being redesigned based on the results of a mediation process.

tion of it might have gone more smoothly had a well-developed regional plan been available. As of this writing, neither case is resolved.

Providing federal backstop authority

Some designs for regional institutions would give authority for siting decisions to a board composed of representatives from the affected states (and perhaps federal and tribal agencies as well). This raises the possibility of internal disagreement; that is, the regional body might be unable to reach a timely decision on whether a proposed transmission project is needed or on the acceptability of a route for the line. To deal with such cases, after a specified time period or under specified conditions,[18] a federal entity could be empowered to rule on the acceptability of the project at the request of the applicant.[19]

**Regional Transmission Planning and
Development of Cooperative Regional Institutions**

Due to the geography of the western U.S., with its comparatively long distances between cities and some of the natural resources used in generating electricity, the western states have gained extensive experience with planning and siting interstate transmission projects. Recently they have begun to develop an institutional framework under the auspices of the Western Governors' Association to aid them in dealing with shared issues related to such projects. Much of this work is being done through a body named the Committee for Regional Electric Power Cooperation (CREPC). CREPC was created in 1984 jointly

---

[18]One possible condition would be the case of a regional transmission project proposed in a state that declines to consider regional costs and benefits.

[19]There are also proposals that would allow applicants to invoke federal backstop authority if a regional entity did not exist and if a state siting agency was not able to make a timely decision about a proposed transmission project.

by the Western Interstate Energy Board, which acts as the energy arm of the Western Governors' Association, and the Western Conference of Public Service Commissions. CREPC has representation from the regulatory commissions, energy agencies, and facility siting agencies in the 11 states and two Canadian provinces in the Western Interconnection. Through CREPC, the western states have begun negotiations to develop a common interstate transmission siting protocol, and are aiming at June 2002 as a target date for a publishable draft.

One of the roadblocks to the formation of comparable institutions in the Eastern Interconnection is the lack of a clear and urgent agenda. That is, without either well-developed regional transmission plans or a collection of actual regional-scale transmission proposals, it is not obvious which states and federal land management agencies need to be talking with each other about what issues. Rather than wait for RTOs to be established and for transmission plans to be developed by them under FERC's direction, an interim approach could be considered. DOE and the FERC could jointly identify key transmission bottlenecks, and FERC could task administrative law judges to work with appropriate parties in each bottleneck area to prepare interim transmission plans. By putting the emphasis on the power of persuasion, such a process would be non-threatening, which would help to elicit constructive responses from stakeholders. The resulting plans would probably flag some important issues affecting groups of states, and thus help to spur the formation of cooperative regional institutions.

The prospect that jurisdiction over a project might pass to a backstop agency after the case proceeds for a certain amount of time could motivate a voting majority of a stalemated regional body to reject the proposal as incomplete before the backstop provision tolls, perhaps in the hope that it would be resubmitted in a form that would win broader support. Further, an agency subject to backstop provisions might be more insistent on the range and detail of alternatives addressed in the initial application, to increase the odds of finding an alternative to which it could say "yes" within the time limit and/or give itself more grounds upon which to declare an application incomplete if necessary.[20] In the end, backstop provisions—linked to time limits keyed to a finding that the application meets a specified standard of completeness—would likely lead to the filing of more complete applications and would impose some discipline on reviewing agencies to act within predictable time periods.

Why Not Just Centralize Transmission Siting Under FERC?

There are obvious challenges in coordinating and harmonizing the views of affected states, local governments, tribal bodies, and federal agencies about proposed transmission facilities. Many observers and industry participants have asked whether it would not be better to enact federal legislation making FERC responsible for transmission siting decisions—particularly because FERC already exercises this function with respect to the siting of natural gas pipelines. Here are some important considerations:

1. Except for areas served by TVA or the federal power marketing administrations, transmission siting is presently a matter of state responsibility. Pre-empting the states and centralizing transmission siting under a federal agency would be a major change, and it is unlikely to win broad acceptance as an appropriate solution to today's siting challenges until less radical measures have been tried and found insufficient.

2. Despite the overarching importance of maintaining the adequacy and reliability of the grid, "all transmission siting is local." Fitting a proposed facility into a landscape where the affected land areas are already used for a wide variety of legitimate purposes will

---

[20]Note that in the case of AEP's controversial Wyoming-Cloverdale proposal, the West Virginia Public Service Commission (which must rule upon an application within 400 days or else it is automatically approved) at one point rejected AEP's application as incomplete and advised AEP not to resubmit its proposal until after the Forest Service had completed its draft environmental impact statement. Resubmitting the proposal would restart the 400-day clock, and the PSC apparently wanted the clock to start after the Forest Service had issued its impact statement.

never be easy. Doing this job well will always require an immense amount of information from local, state, and regional sources, as well as consultation and negotiation with and among many of these parties. Transferring transmission siting responsibility to a single federal agency could mean over-centralization, resulting in delays, hasty, or poor decisions, or all three.

3. The existing process for siting natural gas pipelines is not necessarily a model to be emulated. Critics emphasize that some pipeline siting cases have also dragged on for years, and assert that the process is not sufficiently predictable. They also complain that most events in the process take place in Washington, D.C., and argue that this imposes a substantial burden on many participants, and effectively precludes participation by others.

4. Improved coordination of federal agency reviews of transmission proposals would continue to be a major concern, even if siting responsibility were centralized at FERC. However, this problem can be addressed without centralization.

5. As indicated in many places in this report, the FERC already faces a long agenda of important and urgent matters related to establishing and maintaining effective competition in the nation's bulk power markets. Many of these matters, in practical terms, can only be addressed by FERC—there is no other credible candidate. In the case of transmission siting, however, the states still want to do the job.

A stronger but much more controversial formulation of the backstop concept that has been proposed by some in the electric industry would be to empower the applicant to appeal to the backstop agency when a reviewing agency acts within the allotted time but rejects the application. This version would be much resisted by the reviewing agencies because it makes the state process appear less important to the ultimate decision on the application. It is also unclear how this structure would actually change the nature of the review process. It might make it more difficult for a reviewer to say "no" to an applicant, knowing that the applicant could turn to the backstop agency for a second opinion, or it might tempt a reviewer to reject a controversial project anticipating that the backstop agency may be more willing to take any political heat associated with approving the project. A pernicious effect on the behavior of applicants could be the emergence, at least in some cases, of "forum shopping." That is, some applicants could become less responsive to the concerns of the reviewing agencies and less willing to spend money to address their concerns, knowing that if they got a rejection they could turn to the backstop agency.

Over time, the criteria and standards used by the backstop agency would tend to become definitive for all reviewing agencies, perhaps making the role and powers of the backstop agency more important than the drafters of the backstop provisions had realized or intended.

Responsiveness to local concerns

A frequent criticism of the regional approach, especially if it is combined with federal backstop authority, is that a regional or federal body will not be sufficiently responsive to local concerns. To address this issue, a regional or federal body could be required to hold extensive local public hearings and weigh the concerns expressed at these hearings against regional and national ones. Historically, many regional federal entities (e.g., the Bonneville Power Administration, the Tennessee Valley Authority, regional offices of the Environmental Protection Agency) have proven to be very responsive to local concerns (sometimes to the consternation of officials in Washington, D.C.).

Should regional bodies be empowered to provide advisory opinions only?

Giving regional entities the power to counsel but not

decide would have the advantage of enabling a panel of experts to provide an objective assessment of a proposed project from a regional perspective without infringing upon the reviewing agencies' powers of decision. The reviewing agencies would be under some pressure to explain decisions not compatible with a regional body's advisory opinion. The disadvantage of this approach is that it further complicates rather than simplifies the institutional landscape for transmission siting. Many parties are strongly opposed to adding new layers to siting procedures or electricity regulation.

Risk of jurisdictional confusion

If a regional siting body were established, states in the area would still likely retain jurisdiction for some new transmission projects, depending on the definition of "regionally significant" used to identify the projects over which the new body would have jurisdiction. If the definition relied on clear empirical criteria (e.g., "all transmission projects of 230 kV or higher"), the jurisdictional boundaries would probably be clear, but there would still be some practical difficulties with the empirical approach. (See the section "Defining 'Regional Transmission Facilities,'" on page E-25, for further discussion.)

Risk of "forum shopping"

An applicant might deliberately design a project to fall into one jurisdictional category rather than the other, e.g., so that the body that the applicant perceived to be most favorably disposed would review the project. This might in some way disserve the public interest. As long as both reviewing bodies are reasonably well conceived and well run (and these are not trivial requirements), the public interest should be adequately served.

NEPA reviews

If a regional body with siting authority included some representatives of federal agencies, this raises the question of what level of federal involvement would trigger the requirement for an environmental assessment or environmental impact statement under the National Environmental Policy Act (NEPA). If a decision by the regional body would supplant the need for an independent review of the project by one or more federal agencies, it seems likely that the regional body would have to conduct an environmental assessment. Depending on the results of the assessment, an impact statement might be required. In general, major new transmission projects usually have significant environmental impacts; in such cases, if federal decisions are required, full environmental impact statements must be prepared.

Alternative Designs for Regional Siting Institutions

There are at least five basic designs that might be considered for regional siting institutions, and many possible hybrids among the basic models. The discussion below focuses on the principal distinctions among the five basic models and is not intended to be exhaustive.

Cooperative agreements

A cooperative agreement would establish a regional entity for the mutual convenience of participating states, tribes, or federal agencies; the participating agencies would not cede any existing authority or responsibility to the regional institution. The regional institution's functions would be limited to activities such as fostering common siting processes and requirements and improving coordination among members to streamline review of regionally significant transmission facilities. Members would probably find it useful to agree on a category of facilities that would fall under the entity's purview, and they would have to agree on how to staff and fund the institution. The parties could begin by establishing a cooperative agreement that would apply only to one specific major case and then decide on the basis of that experience whether to continue to proceed case by case or to establish a standing agreement.

Interstate compacts

An interstate compact is an agreement among or between states to establish an institution that has the power to act for all of them in a specific area. Establishing an interstate compact is a complex process, especially if more than a few states are involved. The legislature of each participating state and the U.S. Congress must approve the compact's founding agreement.

For a compact on transmission siting, many states might have to enact legislation to authorize their public utility commissions (PUCs) to cede specific authority to the regional body or to share authority or provide guidance to the commission concerning the circumstances under which it should defer to the regional body. The founding agreement would have to define the class of transmission facilities that would be subject to the commission's jurisdiction and establish how the commission would be staffed and funded.

Agreements for compacts typically specify that the governors of the participating states will appoint the compact's commissioners. Voting representation on a compact commission tends to be controversial because of differences in the sizes of states and how to set each state's share (e.g., based on population or contribution to gross domestic product) as well as the likelihood that some states would probably be more affected by the commission's activities than others. Smaller states tend to prefer one-state, one-vote structures so as not to be overruled by larger states.

Interstate compacts have been established for many purposes, and some have been much more successful than others. They ultimately depend on cooperation and goodwill among the member states. If states are strongly at odds on an issue, a compact commission may find it difficult to solve the problem. Conceivably, a provision for federal backstop authority could be included in the founding agreement to deal with potential stalemates.

Another possible problem with the compact model in the current context is that federal agencies are not subject to interstate compacts. Cooperative agreements could be devised between a compact commission and appropriate federal agencies, but the arrangement would be comparatively informal. Another question is whether the founding agreement could be fashioned to facilitate participation by Native American tribes.

Independent regional entities

The independent regional entity model offers considerable flexibility (regional authorities have been established through federal legislation to address a wide range of problems).[21] Affected agencies (state, tribal, or federal) would have to agree on a conceptual design for a regional authority that would accomplish their common purposes, and then appropriate federal legislation would have to be crafted and enacted to serve those purposes. This approach requires the support of most of the affected states, but it is significantly less formal than the process for establishing an interstate compact.

Presumably, a board of commissioners would head a regional authority, and the enabling statute would set the

---

[21]This model probably comes closest to accommodating the intent of the Task Force on Electric System Reliability to the Secretary of Energy Advisory Board in its Recommendation #25. (See DOE 1998).

criteria for appointment to the board. One approach would be to use the siting boards that currently exist in some states as a model, with commissioners from relevant state and federal agencies or tribal institutions nominated by governors, tribal authorities, or the President.[22] Thus, this model accommodates federal participation more readily than an interstate compact. The designers of the new entity would have to decide how best to balance federal and state interests, particularly with respect to voting powers and whether there would be federal backstop authority.

As with the other models, designers would have to define the class of transmission facilities subject to the new entity's jurisdiction and establish a funding mechanism. If the new entity were to have final siting authority on behalf of federal agencies, the enabling legislation would have to amend the enabling laws for those agencies. Similarly, state legislatures would have to make appropriate changes to their respective siting laws. A sunset provision could be included to ensure future review of the need for and effectiveness of the new entity.

Joint federal-state boards

Although there are precedents in the telecommunications sector for the establishment of joint regulatory boards, this model has not been used in electricity regulation despite periodic expressions of interest by the National Association of Regulatory Commissioners (NARUC) and various states. Further, the relevance of this structure to transmission siting, at least under existing law, is at best uncertain.

Section 209(a) of the Federal Power Act authorizes FERC

---

[22]A critical design element would the process for removal of commissioners from the regional board. Serving at the pleasure of the appointing authority is quite different from serving for a distinct term, for example. Another significant matter to address is how such an organization would be staffed.

to refer an electricity matter under its jurisdiction to a joint state board composed of nominees selected by the respective state utility commissions or by the state's governor if there is no state commission. A joint board is to have the same power, duties, and liabilities as a commissioner at FERC who has been directed by FERC to hold hearings. Thus, a joint board for an electricity matter, assuming unanimity among its members, would be equivalent to a sixth commissioner at FERC with respect to FERC decisions on the matter.[23]

However, under current law, FERC has no jurisdiction over transmission siting, so it would have no basis upon which to call for the establishment of a joint board to address transmission siting issues.

Regional FERC offices

FERC could be directed through federal legislation to

---

[25]The full text of Section 209(a) reads:

[FERC] may refer any matter arising in the administration of this Part to a board to be composed of a member or members, as determined by the Commission, from the State or each of the States affected or to be affected by such matter. Any such board shall be vested with the same power and be subject to the same duties and liabilities as in the case of a member of the Commission when designated by the Commission to hold any hearings. The action of such board shall have such force and effect and its proceedings shall be conducted in such manner as the Commission shall by regulations prescribe. The board shall be appointed by the Commission from persons nominated by the State commission of each state affected, or by the Governor of such State if there is no State commission. Each State affected shall be entitled to the same number of representatives on the board unless the nominating power of such state waives such right. The Commission shall have discretion to reject the nominee from any State, but shall thereupon invite a new nomination from that state. The members of a board shall receive such allowances for expenses as the Commission shall provide. The Commission may, when in its discretion sufficient reason exists therefore, revoke any reference to such a board.

establish offices in each RTO's area; each office could be made responsible for transmission siting and rate regulation within the region. Such legislation could limit FERC's regional activities to matters such as hearings before administrative law judges and staff reviews of siting applications and could reserve final decision authority to the commission. The legislation could also direct FERC regarding the creation of regional joint state boards on transmission siting, the weight to be given to decisions by such boards, and how FERC's siting decisions should take into account the views and expertise of other federal agencies and Native American tribes.

DEFINING "REGIONAL TRANSMISSION FACILITIES"

If regional transmission siting entities were established, the category of facilities subject to the jurisdiction of these bodies would probably need to be defined. The subsections below address possible criteria for this definition and the institutional context in which they might be applied.

Objective Criteria

One way to define the transmission facilities that would fall under the jurisdiction of a regional siting body is to use objective indices, such line voltage or length or whether the line would cross state boundaries. Although these criteria may sound reasonable, they may not always yield the expected results. For example, in some sparsely populated areas, lines that serve transmission functions may be comparatively low voltage; conversely, in some densely populated areas, distribution lines may be designed for economic reasons to operate at high voltages. Another example is that a facility may be used in part for transmission and in part for distribution purposes. One way to deal with problems of this kind is to create a definition based on objective criteria with a mechanism that would allow an affected party to petition for a waiver, based on demonstrating that the criteria should not be applied in a specific case.[24]

Functional Tests

An alternative for defining the jurisdiction of a regional body is to apply functional tests that gauge whether a facility would be used primarily or wholly for transmission and define the degree of its expected contribution to the reliability of the regional grid. A significant objection to this approach is its lack of transparency—applying it could require hearing and evaluating evidence before a decision could be made about whether a proposed facility is regionally significant.

Economic Test

An economic test could be devised to estimate the probable economic benefits that a line would provide for consumers over a given period through either improved access to lower-cost generation or mitigation of potential market power. This estimate could be compared to an agreed-upon threshold for determining regionally significant projects. This approach might also require gathering and evaluating evidence.

In short, there are no easy, straightforward criteria. However, determining the criteria would be more important in some institutional contexts than others. For ex-

---

[24]Many states currently use objective criteria (such as voltage and line length) to determine whether transmission distribution projects need state approval. Projects that do not meet the defined threshold in these states still have to meet local zoning, safety, and other requirements, but they do not have to go through the full state siting review process.

ample, if the institution's principal function is to facilitate cooperation among the reviewing agencies in the region, if the agencies retain their existing authority, and if no federal backstop mechanism is established, then no jurisdictional changes would result from the designation of a project as a "regional project." A "regional project" would be channeled through the regional cooperative process, but no other changes would ensue. As a result, the criteria for determining a regional project would be less important (and less likely to be the focus of litigation) than if designation as "regional" would mean that a project might under certain conditions be shifted onto a federal jurisdictional track. As long as jurisdiction would not be affected, the most important choice the reviewing agencies would have to make could be whether they wanted to channel all transmission projects through the regional body, or only a subset of projects deemed to have regional significance.

By contrast, if the regional institution was given the power to decide siting questions, the scope of its jurisdiction would be much more important, and the founding parties would probably wish to define criteria for jurisdiction very carefully. Similarly, if a federal backstop mechanism were created by federal legislation, the legislation would probably have to address jurisdiction. One approach would be to sidestep the criteria altogether and specify that under certain conditions (e.g., failure of a reviewing agency to act within a specified period, or rejection by a reviewing agency of an RTO-approved transmission project), the applicant could petition the backstop agency to take the case. Another alternative would be for the legislation to direct the backstop agency to conduct a rulemaking procedure to establish appropriate criteria for identifying transmission projects of regional or national importance.

IMPROVING THE EXISTING
STATE-BASED SITING PROCESS

Regardless of how the debate evolves over whether regional or federal authorities should be responsible for certain aspects of transmission siting, states will continue to be responsible for siting a large proportion of the nation's new transmission facilities. Thus, it is worthwhile to consider how the state-based siting process could be improved.

Transmission proposals fall typically into one or more of three categories:

- Those needed to connect a new generator to the grid,

- Those needed to meet reliability standards, and

- Those needed to enable increased electricity trade.

Some projects are very small in geographic scope; others extend for hundreds or even thousands of miles.[25]

Although there is debate about the scope of possible federal or regional responsibilities for transmission siting, state authorities will continue to review dozens of transmission or transmission-related proposals each year, and responsibility for siting generation is likely to remain with the states. Similarly, most legislative proposals that would shift some jurisdiction for transmission siting away from states (e.g., transfer a "backstop" authority to FERC) nonetheless leave states with the primary authority for this function. State-based transmission siting processes vary considerably across the U.S., and, for the most part, worthy projects are approved, and deficient projects are discouraged, improved,

---

[25]A recent proposal (not yet filed at a siting authority) would build approximately 2,000 miles of transmission lines to connect new coal generation in Wyoming with load centers in Chicago and Los Angeles.

or rejected. Most transmission projects are intrastate and small in scale.

Even successful siting cases may have shortcomings, and some cases illustrate recurrent criticisms of state-based transmission siting that warrant attention. Some observers believe that the cases that could have been handled better represent exceptions to a basically sound system. Others see these cases as symptomatic of a need for fundamental changes.

Accountability

Any system of regulation must have and retain public confidence. Generally, regulators earn public confidence by being fair, competent, and consistent over time. In the United States, the general practice is to assign responsibility for regulation to the level of government that can most effectively serve and protect the interests of the citizens affected. This practice allows local conditions and differences to be reflected in regulatory decisions, and non-local considerations can be taken into account when appropriate.

Improvements to Siting Processes

State laws governing transmission siting are the product of serious debate among elected officials. Likewise, state siting decisions are the products of a careful weighing of evidence in light of public policy expressed in statutes. Although state siting laws and processes have been conscientiously developed, improvements may be needed to maintain a reliable and adequate electricity grid. Some possible changes are discussed below.

"One-stop" siting process

Some states place the authority for considering transmission siting proposals in a single agency, which may be the state regulatory utility commission or a siting board made up of decision makers from several government departments. This structure makes accountability for siting decisions clear, and it enables applicants to become familiar with a single process. If local authorities have a role in the approval process, it is important that the state be able to impose on all local reviewers a common, statewide perspective regarding the regulated utility system.[26]

Interstate projects would be eligible for one-stop treatment only if the affected states combined their efforts into a regional siting process. This principle has many supporters, but the procedural requirements would be very demanding; the authors are aware of no successful attempt at a voluntary, one-stop, multi-state siting process.[27] The dilemma for states is often thought to be whether the state siting authority should focus exclusively on protecting the state's interests or should take an expansive view and consider regional interests. This is a false choice. The long-term interests of most consumers are best served by addressing regional grid needs while accounting for state interests at the same time.[28]

---

[26]For example, recent legislation passed in Colorado modified the standing of local authorities in transmission siting matters. The PUC can now pre-empt the decision of local authorities if there is a compelling state interest.

[27]Ohio has a statute that explicitly authorizes its transmission siting authority to cooperate with other states, but this process has yet to be tested. The western states have begun negotiations concerning a common interstate siting protocol for the west, which could result in something like a regional one-stop process.

[28]There are many examples of state siting orders that make a special effort to acknowledge the importance of regional concerns. There are also examples that do the opposite.

Three Views of State-Based Siting

Three organizations with distinctly different perspectives about U.S. electricity policy are the Western Governors' Association, the Edison Electric Institute, and the Electricity Consumers Alliance. Although many parties have views about how to change the transmission siting process, the views of these organizations illustrate that there is a broad range of opinions.

The Western Governors' Association (WGA) is a policy forum serving 18 western states, including Alaska and Hawaii. The organization has a long-standing interest in transmission siting and energy policy. The WGA position is:

- Transmission expansion should support three key priorities: enhance reliability, reduce consumers' costs, and promote fuel source diversity.

- Need should be established using regional criteria.

- Siting should remain the responsibility of the states.

- The states should collaborate in the review of interstate transmission projects, and federal land management agencies should join this collaboration.

The Edison Electric Institute (EEI) is a trade association that represents the interests of investor-owned electric utilities. The EEI position is:

- States should have a limited amount of time to review any transmission project.

- If a state rejects a project or does not rule within the allotted time, FERC should be authorized to take the case as it stands and rule upon it within a specific time period.

- Other EEI recommendations concerning federal land management agencies focus on enhancing coordination and attention to deadlines in agency reviews of siting proposals.

The Electric Consumers Alliance (ECA) addresses electricity policy issues nationally and in key states on behalf of small consumers and their local organizations. The ECA position is:

- Determination of need for new transmission should be made by a regional transmission organization.

- Federal, state, and local reviews should take no more than 12 to 18 months.

- Reviews by more than one agency within a single state should be combined. Similarly, reviews by more than one federal agency should be combined.

- If federal or state reviews are not complete after the allotted time, FERC should take and rule on the case.

- The rights of individuals must be respected in the siting process.

Sources: The Western Governors' Association published its views in Conceptual Plans for Electricity Transmission in the West, 2001). The Edison Electric Institute (EEI) is a trade association for investor-owned electric utilities. EEI's views were conveyed to the authors in a personal conversation with Rich Loughery and Henry Bartholomew. The Electric Consumers Alliance (ECA) represents hundreds of rural, senior, low-income, small-business, minority and other consumer organizations. ECA conveyed its views at a DOE public hearing on September 28, 2001.

States will also need to address the allocation of costs for a regionally justified transmission project. An RTO or a tightly operated ISO[29] will administer this matter once the project is built but typically does not have an active role at the project review stage. If there is a problem with the allocation of costs and benefits among states and their consumers, these money matters should be negotiated under

[29]A tightly operated pool is one that controls and dispatches all the generators to reduce overall costs, and internalizes numerous cost allocation decisions in its rate structure.

pressure from regulators (as they often are in natural gas open-season proceedings).[30] Siting authorities can send signals to developers and allow reasonable time for proposals to be adjusted to address such concerns. Authorities can also encourage project planners to address this subject with stakeholders and the public before an application is filed.

Ex parte rules control how information flows to and from the regulatory body; they are intended ensure a fair process free of abuses by parties who have ready access to decision makers. The evidentiary basis for an order should be clear from the record. However, ex parte rules can hinder the management of siting dockets and negotiations with the applicant or other reviewing agencies by shielding the siting authority from valuable insights more likely to emerge in conversation than in cross-examination. Beyond speaking through their orders, regulators can find ways to communicate constructive information in a fair way, using methods such as workshops, special masters and other alternative dispute resolution methods, written questions to the parties, status orders, etc.

Maximum time limits

Most transmission siting proposals are small in scale and are reviewed and acted on by the relevant state authority within a year. Larger projects attract more attention from intervenors, are more complex, and may take longer. In some protracted cases, the siting authority may, because of reluctance to reject a project that appears to have merit but needs modification before it can be approved, allow the applicant time to correct deficiencies that emerge during the proceeding.

---

[30]The April 2001 Connecticut Siting Council decision to reject the Cross Sound Cable project included a warning, presumably directed at successor proposals regarding the allocation of costs as compared to the expected benefits.

In general, however, siting authorities should strive to maintain schedules and avoid delays. Among other things, this means not allowing opponents of a project to hold up the process. Opponents must have a fair opportunity to gather information and present a case but should not be allowed to take control of the calendar. The project proponent can help prevent this kind of delay by presenting a credible array of alternatives so that opposing parties cannot obstruct proceedings by calling for inquiries into reasonable alternatives that have not been addressed in the proposal.[31]

As an alternative to allowing the siting review calendar to be based on judgment calls, some states impose a time limit on the process. However, if a time limit is to have a positive effect, the time allowed must be sufficient for a review that will meet public expectations for thoroughness and fairness. A very tight time limit can too frequently put the authority in the difficult position of nearing the deadline with inadequate evidence to find in favor of a project. A system that frequently results in rejections on procedural grounds or approvals by default is not a good system.

A recurrent complaint from prospective applicants is that siting processes without time limits are too unpredictable. For example, unpredictable time frames can negatively affect project financing; an applicant may be reluctant to spend the money to develop a proposal and support it through the approval process unless it is reasonably certain that it will be able to obtain financing for the construction phase of the project. However, potential financial backers may be unwilling or unable to address the financial details of a project if they do not know when construction might

---

[31]Of course, if there is a superior alternative, the process must accommodate it. Proponents of transmission projects should do their best to ensure that there are no superior alternatives and expect the review process to ratify that view.

begin or be completed, and a project tied up in a protracted review is more likely to be adversely affected by ongoing changes in bulk power markets. Reasonable time limits on transmission siting processes would help dispel the uncertainty that appears to hamper many business decisions in the transmission sector.[32]

Clarify approval criteria

Fortunately, many transmission proposals that come before siting authorities address unambiguous needs to improve reliability or to respond to growth. The difficult cases are ones in which the facts do not line up well with the approval criteria, or the criteria themselves are inadequate for the specific situation. States should examine the approval criteria in their siting statutes in light of the significant changes occurring in bulk power markets (see "The Regional Perspective," on page E-13). In addition, when a case exposes a weakness in the statute, this should be addressed by the state legislature as soon as possible.[33]

Cost recovery rules and grid investment needs

Utility costs cannot be recovered from consumers without rate proceedings. Many utilities' rates are frozen or capped for long periods as part of a regulatory agreement, as imposed by a legislature in electric restructuring laws, or for punitive reasons. Without performance incentives or the opportunity to recover extraordinary costs, a utility may decide to avoid major investments even when they are

---

[32]A complete proposal, based on standards established by statute and rule, is key to making a time limit work. Until a proposal is complete, the "clock" should not start.

[33]Legislators are sometimes reluctant to "open up" a statute for fear that others will take the opportunity to press for other changes. This concern must be balanced against the need to update an important process.

needed. When considering rate freezes and caps, regulators and legislators should consider the horizon of prospective utility investments and consider whether a cap will stifle important projects.[34]

Federal incentives for state changes

In some instances, state siting processes based on an accumulation of law and precedent may no longer be adequate to address the challenges associated with the current restructuring of the U.S. electricity industry.[35]

Given the arcane nature of transmission siting and the potentially difficult political challenge of updating the siting process, the federal government may be able to facilitate needed change by means of incentives.

Federal sponsorship of workshops and development of model legislation are worthwhile approaches; another initiative that has significant support would put the Federal Energy Regulatory Commission in a backstop role to state siting authorities. This approach, which would require changes in federal law, would give FERC siting jurisdiction over proposed "regional transmission facilities" (See section on "Defining 'Regional Transmission Facilities,'" above) if

---

[34]Utilizing traditional regulatory tools like Construction Work in Progress accounts or simply booking and deferring costs for future regulatory treatment can provide utilities with assurance that they will recover the costs of needed transmission investment incurred during a rate cap, including a reasonable return on investment after the end of the rate cap. However, if the cap is part of a performance ratemaking plan, and the utility has accepted the risk that such costs may be needed during the period of the plan, then asset depreciation would start normally, and the utility could include the depreciated costs in the consideration of post-plan rates. In this latter case, utilities would still have incentives to pursue cost-effective transmission investments because efficiency improvements inure, at least in part, to the utility's profits in performance based ratemaking.

[35]This subject requires extensive analysis and lends itself to the "best practices" project discussed below in "Federal Assistance."

affected states fail to act within a specified period.[36] Many observers expect that if FERC had this role, most states would intensify and coordinate their efforts and complete reviews in time to avoid an unwanted change of venue to the backstop authority.

An approach that some observers find less aggressive would be for federal law to support or assist the formation of cooperative regional bodies composed of officials from affected states; these regional bodies could be convened to coordinate the review of regionally significant transmission proposals. (This idea is explored in the section "The Regional Perspective," above.) These regional institutions could be aided by findings of need from the soon-to-be-formed RTOs. The question of what authority states should retain in future siting processes is currently stalemated between advocates of state authority and proponents of federal authority.

IMPROVING AGENCY AND INDUSTRY PRACTICES

Not all barriers to siting of new transmission lines are related to the state-based review process. Some delays and rejections result from omissions or other types of problems with transmission proposals or with the practices of transmission owners.

The subsections below address changes in practice by prospective transmission siting applicants that could improve the quality of regulatory outcomes. This section also

---

[36]FERC backstop authority could also be exercised if state siting authorities addressing a regionally important multistate project disagree on whether the project should be permitted. This is different from a trigger based on a time deadline because in this case the states would have executed their responsibilities. FERC could determine whether some compromise or blending of interests among the affected states would be possible.

turns attention to the federal government, addressing siting on federal lands, siting by federal utilities, and other actions the federal government can take to improve siting results.

The subjects in this section are linked by improving methods, utilizing existing methods better, more effectively deploying new methods, and communicating among all affected parties more effectively. A positive outcome would be one in which the transmission owners' interest and the public interest are better aligned than they appear to be today.

Effective Presentation of Alternatives

Transmission siting proposals are complex, especially for large-scale projects designed to improve reliability or enable increased energy transfers over wide regions. To aid decision makers in making a sound choice about whether to permit a project (and to prevent critics from derailing a project by shifting attention to other options), a proposal should include a detailed presentation of the alternatives considered.[37]

Alternatives enhance credibility and public confidence

A proposal that presents and compares alternatives shows that the proponent is focused on meeting a system need in the best way, not on getting a particular project built. Addressing alternatives shows the applicant's confidence that the proposal represents the best approach to meeting a system need. This approach can be aided by undertaking an open planning process once a need has been recognized but before a solution is selected; the public should be engaged in

---

[37]This is not usually a concern for transmission that will interconnect a generator with the grid.

this process to assist the transmission company in combining its own and public interest priorities in the decision process.[38] This process improvement should not be used, however, as a way to shift the responsibility to develop alternatives to intervenors. Many permitting agencies already require that proposals include alternatives. Agencies that do not should consider adding this requirement as an investment to speed the overall process.

Range of alternatives must be broad

Even when an applicant presents alternatives, the range addressed may be too narrow. Efforts to define a generic list of alternatives that should be addressed are difficult because of the inherent variety of grid needs and circumstances. Instead of mechanically addressing a list of required alternatives, an applicant will likely fare better by determining what alternative routes or alternatives to transmission are likely to be considered relevant by the regulators and potential intervenors and addressing these options in detail. (The applicant will readily learn about these alternatives during a transparent planning process.)

If important alternatives are not evaluated in the proposal, they are likely to be introduced by public advocates or other intervenors who may assert that the alternatives represent a better approach than the proposed project.[39] It is also worth adding that a transmission line serves no other pur-

---

[38]Southwestern Public Service, then a subsidiary of New Century Energies, conducted such an open process in building a transmission line in Kansas. As a result the Kansas Corporation Commission approved the segment of the project in its state despite the lack of direct and immediate benefit to Kansas. (Personal communications with Mark Doljac, Kansas Corporation Commission.)

[39]An example is a transmission project in New Mexico that was rejected after local generation and efficiency alternatives were proposed by the state Attorney General and other intervenors.

pose than to conduct power, but other options such as increasing energy efficiency, managing load, and constructing local generation, may have distinct, positive externalities in the community while also contributing to reliability. Franchised wires companies are usually concerned with the general economic well-being of their service areas, so they have reason to consider a broad range of potentially beneficial local investments.

Advantages of Open Planning

A frustration that is sometimes expressed in the midst of a transmission siting dispute goes something like this: "If only the applicant had spoken with us before going public with the proposal. Now both sides are digging in for a fight." Costly proposals to build new lines sometimes seem to come out of the blue because needs are not articulated ahead of time, if ever; once a transmission corridor is proposed, land owners and other interested parties may feel as if set upon by a powerful force. It does not have to be this way. Although some parties will oppose power line proposals regardless of the circumstances, others may be moved to oppose a project not so much because of its content but because of perceptions that the proponent is behaving in an arrogant or paternalistic fashion or making a unilateral decision. Despite the costs of regular reports to the public about the state of the transmission grid and its expected needs, it is in the interest of both the public and the applicant or RTO to make these reports. System needs can be tracked as they evolve from technical indications into demonstrable problems. Discussions about how to address growing concerns can be particularly productive if they involve affected parties and all relevant information is available to anyone who cares to look for it. Early identification of potential problem areas also allows small-scale responses like distributed resources the best opportunity to contribute efficiently to a solution.[40]

Deterministic and Probabilistic Planning

Deterministic analysis identifies possible events (e.g., failure of a large generator) and studies their effects on reliability. The analyst assesses the likelihood of these events based on professional judgment. Probabilistic analysis uses a rigorous statistical method to assess the likelihood of an event and its effects. Probabilistic analysis allows for relatively easy numerical comparisons of alternatives, but these comparisons may seem more precise than they actually are because the results are highly dependent on the quality of forecasts of future equipment performance. Deterministic approaches are more traditional and less costly. Both methods are valuable. Regulators should encourage the use of both so decision makers can have the most complete information possible.

Impact of Rate Design on Decision Making

As its participants know well, there are many ways to regulate the electric utility industry. The rules and rate designs in force at any given time affect the decisions and behavior of the players. Some examples follow showing the effects of rate design on the assessment of new transmission proposals:

- If the cost of a new transmission project is "rolled in" to average regional transmission rates, the new transmis-

---

[40]A related topic is that the grid in which investments are made today will not be the same grid in just a few years. Loads will change, new generation will be built, and some units may be retired. One merit of a transparent process is that it helps focus on the investments that are most likely to make sense for a wide variety of futures.

sion will be far easier to justify than if the same costs are assigned only to the group of consumers in the region whose changes in electricity usage have caused the investment to be necessary.[41]

- If the costs of an alternative are treated as rolled in while the costs of competing alternatives are charged incrementally to those whose energy use has caused the need for the new transmission, the utility will tend to select the alternative whose costs are rolled in even if it is more expensive and less effective at meeting grid needs.

These are not hypothetical examples. The first case is typical in the New England Power Pool (NEPOOL), where the cost of "pool transmission facilities" is borne by all consumers in New England. Although these facilities are not intended as local interconnection service and are in principle necessary for reliability, their need is often the result of demand growth in a distinct part of the whole region. Nonetheless, everyone pays. The second case is typical in most regions. Distributed resources such as energy efficiency and local generation are the best answers to some grid problems.

---

[41]A corollary to this idea is drawn from experience with highways. If new roadways are built to address congestion without addressing lower-cost ways to reduce traffic, and if the source of the demand for the new roadways does not pay the cost for the new construction, the new road can generate more traffic. That is, more traffic than expected will use the new roadway because it is available, and congestion will increase more rapidly than highway planners would have predicted based on prior patterns. Similarly if a new remedial connection to the grid is built and the costs are assigned to society rather than to the connection's direct beneficiaries, the connection can result in increased demand (either from inefficient generation siting or even greater volumes of long distance energy trading) and therefore increased congestion. Some would call this an implicit subsidy. The result of this scenario is increased congestion, much more rapidly than would be expected based on prior patterns.

Yet the system-wide financial support available for transmission to assist the grid is not available for these competing alternatives. Basic economics suggests that when the cause of an investment can be clearly be assigned to a specific group of customers, those customers should pay for it. Implementation of this rule by regulators is complex in practice though congestion transmission pricing is a very positive step in this direction. Ignoring this rule will adversely affect the nature and efficiency of future utility investments.[42]

Encouraging Innovation

One way that the transmission siting process can be improved is for regulators to reward applicants for bringing forward innovative ways to address transmission grid needs. There is evidence of this already, as DC proposals, undersea projects, and flexible AC transmission system (FACTS) devices begin to appear on grid expansion plans. Industry and DOE should continue their attention to the pace and direction of transmission- related research and development, and the industry should continue to educate regulators about the merits of new approaches and devices that can enhance the grid.

Effects of Cost Minimization

Some parties are critical of existing regulation because returns on equity investment are thought to be inadequate compared with the risks of the enterprise and the value added by transmission facilities. In this view, transmission costs are roughly 10 percent of retail electric rates; a modest

---

[42]This idea can be extended to the retail regime as well. The State of Connecticut directs system benefit funds to support demand-response programs in designated transmission- and distribution-constrained areas. (Also, see Moskovitz, 2001.)

increase over this figure should be acceptable to consumers if the result is greater incentive to propose needed projects. Allowing higher proposal costs would also tend to widen the range of economically competitive alternatives.

At the same time, applicants sometimes resist adding features to their projects that would increase costs but bring the proposals in line with public policy concerns. Examples of such features include:

- Selective undergrounding,

- More attractive tower designs and wire placements,

- Longer routes around sensitive areas,[43]

- Zigzag corridors as an alternative to long, straight wooded corridors, and

- Sharing of more financial benefits with affected landowners.[44]

Some might suggest that these elements "gold plate" a project. Others see these features as real costs necessary to win support and fit a needed project into surroundings that are not blank slates but lands protected by legitimate property rights and valued by society. A transparent planning process that focuses more broadly on addressing future needs will aid applicants in identifying beneficial improvements to budding projects.

---

[43]See description of Cross Sound Connector project in the section "Two Instructive Transmission Siting Cases," on page E-8.

[44]Utilities express concern that premature identification of a route may result in increased easement costs. In contrast, rumors of a prospective transmission project may adversely affect land values and burden landowners with uncertainty. We suggest putting all the facts on the table and relying on the siting authority (and courts if necessary) to rule expeditiously on the project and its route and to set fair and reasonable easement costs.

Need for Complete Applications

Transmission siting is a difficult process at best. When a proposal is incomplete, the process becomes still more difficult. The reasons for incomplete applications range from a lack of familiarity with the rules and expectations of the siting authority to intentional omission of significant information. In any case, the burden is on the applicant to know and abide by the spirit of the rules. This is not just an issue of fair play; trust is a fragile commodity in a process where the threat of eminent domain always looms, even though it is rarely mentioned and even more rarely used. When applicants do not abide by the rules of the process, they may lose the trust of the public and the siting agency. Once trust has been compromised, it is difficult for a review process to reach an outcome that will be in the public interest and be so recognized by most parties.

Transmission Company Perceptions of the Siting Process

In some jurisdictions, there is anecdotal evidence that at least some transmission system problems are not being addressed because utility executives are concerned about the hostile reception they expect that proposals would receive from the state siting process.[45] Utilities holding this view assume they would lose in the court of public opinion and waste financial and human resources in the attempt. It is difficult to evaluate these anecdotes for several reasons. A utility speaking freely and acknowledging reluctance would risk a regulatory ruling that it had been imprudent for failing to pursue construction of needed facilities. Further, the root cause of the reluctance may relate to factors other than the siting process. The existence of these stories, however, is

---

[45]Other possible factors include uncertainties regarding cost recovery in a state or how costs would be allocated among states and companies for interstate projects. Local politics may also be a factor.

clear indication of a problem. One objective of reform of the siting process should be to ensure that the process is perceived as welcoming good proposals and offering a fair test to all projects. A process in which utilities with an obligation to deliver are so intimidated that worthwhile projects remain under wraps does not serve the public interest.

Solving Existing Aesthetic Problems in Combination with New Transmission Projects

In some cases, a new transmission project can provide the means to resolving a community's existing aesthetic problem. Consider the case of an aging industrial waterfront area that has the potential to be transformed into a civic and tourist center, but its best views are marred by an accumulation of high voltage lines left over from its industrial past. Some communities are working with their utilities on such projects by finding ways to remove some or all of these lines in conjunction with upgrading other transmission lines nearby. This somewhat radical approach—removing still-functional facilities from service for aesthetic reasons—can produce a more efficient transmission system, while strengthening public support for an otherwise intrusive project.

One example is in Minnesota. As part of the controversial Chisago-Apple River proposal, a mediation process revealed the existence of an opportunity to clean up the visual effect of accumulated power lines in the city of Taylor Falls, MN. Power lines would be removed, and one 161 kV line would cross the river in its place. The concept would also place the new line underground for some distance near the waterfront. Execution of this idea is still pending; Xcel and Dairyland Cooperative have not yet filed the new proposal with siting authorities in Minnesota and Wisconsin.

Another example is in Vermont, where the Vermont Electric Power Company and the City of Burlington are working together in advance of a major VELCO transmission siting proposal to see if lines on the redeveloped waterfront of Vermont's largest city can be removed as part of the project. Advance planning ensures that regardless of the decision, all sides will know that great effort was made by VELCO to find positive collateral benefits.

Federal Actions to Improve the Siting Process

There are several ways, described in the subsections below, that the federal government could promote improved transmission siting performance in the United States, independent of how jurisdiction is apportioned between state and federal regulators.

Improving federal land management agency reviews

Probably the second-most-often-heard category of complaints about the transmission siting process (after concerns about the state process) relates to federal land management agency reviews of proposals. Almost 29 percent of the total land area of the United States is owned by the federal government and managed by the Departments of Defense, Agriculture, Interior, and other agencies (Statistical Abstract of the United States, 2000; see box below for additional details). In addition, other non-federal land areas such as airsheds, wetlands, navigable waterways, and coastal zones are subject to federal oversight by the Environmental Protection Agency, the Corps of Engineers and other agencies.

These complaints fall into four general categories:

- There is often inconsistency within an agency in the ways local or regional land managers review transmission projects.

- When two (or more) federal agencies are involved, there is frequently inadequate communication and coordination between them.

- Review of transmission proposals does not appear to be important in comparison to the primary mission of the agency.

- Federal agencies frequently wait to conduct their reviews until state reviews are completed and a final route has been selected. This introduces the risk that a federal agency may require a route change, leading to another (time- and cost-consuming) iteration in the state process.

(See box on Alturas case (next page), which illustrates some of these problems.) It should be noted that research for this paper also found reports of good cooperation between states and federal agencies.

Distribution of Federal Lands in the United States

Although almost 29 percent of the land area of the United States is federally owned, the distribution of this land is very uneven. Nearly 38 percent of all federal land is in Alaska where almost 68 percent of the state is federally owned. Another 54 percent of all federal land is concentrated in the 11 states of the contiguous U.S. that are located wholly or partially west of the Continental Divide. Additional details about these 11 states are presented in the following table:

State	Total Area (Acres, in 000's)	% Federal Land
Arizona	72,688	45.6
California	100,207	44.9
Colorado	66,486	36.4
Idaho	52,933	62.5
Montana	93,271	28.0
New Mexico	77,766	34.2
Nevada	70,264	83.1
Oregon	61,599	52.6
Utah	52,697	64.5
Washington	42.694	28.5
Wyoming	62,343	49.9

Source: Statistical Abstract of the United States, 2000 (U.S. Dept. of Commerce, December, 2000),Table No. 381 (1997 data).

The Alturas 345 kV Intertie Project

This project demonstrates some reasons why potential developers of transmission facilities regard gaining permits from affected federal agencies as one of the most difficult and frustrating aspects of transmission siting.

The Alturas line is 163 miles long and runs between Reno, Nevada, and Alturas, California. About 20 miles of the line is in Nevada and the balance is in northern California. The line was needed primarily to support reliability in the fast-growing area around Reno, and to enable the applicant, Sierra Pacific, to gain access to low-cost hydro from the Pacific Northwest for the benefit of retail customers in both Nevada and California.

The project was proposed to the Nevada Public Service Commission early in 1993 and the Commission approved it in November 1993. Sierra Pacific then turned to the other affected agencies: the California Public Utilities Commission (CPUC), and several federal agencies [the Bureau of Land Management (BLM), the U.S. Forest Service, the Bonneville Power Administration (BPA), and the U.S. Fish and Wildlife Service (FWS)]. BLM became as the lead federal agency for the purposes of preparing an environmental impact statement because it had the most affected acreage. The Forest Service had two affected areas, three line miles in the Modoc National Forest in California, and eight line miles in the Humboldt-Toiyabe National Forest in Nevada. The California Public Utilities Commission became the lead agency for state environmental purposes.

In the spring of 1994 BLM and CPUC jointly hired a consulting firm to prepare an environmental impact report (EIR) for the state and an environmental impact statement (EIS) for the federal agencies. The applicant paid the cost of this work. The draft statements were issued for comment in March 1995. In the fall of 1995, the applicant believed that the comments received could be satisfactorily addressed through several kinds of mitigating measures. BLM issued the final EIS in November 1995, and approved its portion of the project in February 1996. The CPUC approved its portion of the line in January of 1996. However, in February 1996 the manager of the Humboldt-Toiyabe National Forest issued a "no action" decision, and argued that the EIS had been flawed because it had not addressed a sufficiently wide range of alternatives, including the alternative of skirting the Humboldt-Toiyabe National

Forest entirely.

The applicant appealed this decision, first to the regional forest manager and then to the deputy chief of the Forest Service. The appeal process took several months, and the results of the appeal were inconclusive. In June 1996 the deputy chief ordered the "no action" decision withdrawn, but he also directed the Humboldt-Toiyabe manager to obtain whatever information was needed to make a new decision. This led to several months of dialogue between the applicant and the Humboldt-Toiyabe manager, and the filing by the applicant of several hundred pages of additional information. The manager of the Modoc National Forest, who had not issued a final decision on the portion of the route that would cross the Modoc area, joined this dialogue.

However, the applicant found that the continuing uncertainty over the acceptability of the Humboldt-Toiyabe route segment was making it difficult to gain required permits from local governments in Nevada that would be needed for the construction phase of the project. These problems led the applicant to examine the option of an alternative route on private land around the Humboldt-Toiyabe National Forest, even though this had several disadvantages. It would put the line into more developed areas, and make it more visible to local residents. This alternative route was about the same length as the initial route, but it was more costly because it would need more expensive towers in several locations, the right of way was more expensive, and additional legal costs would be involved. At length Sierra Pacific decided to pursue the private-land route and withdrew its application to cross the Humboldt-Toiyabe area in February 1997. Due to these route changes, the applicant had to go through some local-level processes a second time in Nevada.

In April 1997, the manager of the Modoc National Forest issued a decision on the EIS, also denying the applicant's request for a permit. Sierra Pacific appealed this decision to the chief of the Forest Service in May 1997, and this led eventually to the issuance of a permit in October 1997. However, several other parties to the proceeding appealed this latter action. After review, the decision to issue the permit was upheld in January 1998.

Construction of the project was begun in February 1998 and completed in December 1998. The applicant estimates that the difficulties with the Forest Service delayed the project by at least two years and led to additional costs of well over $20 million.

Addressing these concerns about federal agency reviews must start with a recognition that a change in priorities is required: applicants deserve a timely, consistent, and substantive response from the federal government. For the same reason that a "one-stop" siting process makes sense at the state and local level, federal agencies should find a way to participate cooperatively and constructively in the overall siting process. This may require additional effort and resources from both the applicant and the agencies to consider alternative routes and solutions earlier in the process.

One option is to centralize individual agency responses to transmission proposals.[46] Special staff groups could be created in the headquarters of appropriate federal agencies to work jointly on reviewing transmission proposals, particularly if efforts to improve coordination among federal agencies and to train and inform regional managers about the importance of the transmission grid do not achieve the desired results.

Another option is to designate a lead agency for cases where two or more federal agencies are affected, and give that agency jurisdiction over all federal matters affected by the transmission proposal. It would be difficult to gain broad support for this approach because it would require some federal agencies at times give jurisdiction to another federal agency regarding land use within their domains. It is worth noting that this approach is not used in siting natural gas pipelines, even though siting such lines is wholly under federal jurisdiction.

A less radical version of this option would be to make the FERC the lead agency for coordinating all federal reviews of proposed transmission facilities, while specifying

---

[46]See comments to the DOE by the Electricity Consumers Alliance, discussed in the section "Improving the Existing State-Based Siting Process" on page E-45.

that other affected federal agencies would participate in the reviews as cooperating agencies, and would retain their existing authorities. Charging one agency with overall coordination of the process, especially one already experienced with environmental and other types of analysis of electricity projects, would help to bring greater consistency and predictability to the federal review process. Further, given FERC's other responsibilities in the electricity area, it would have stronger reasons than most other agencies to press for good coordination, and eventually it would also have regional transmission plans at its disposal to use in confirming whether a proposed transmission line is needed. Presumably, establishing this approach would require federal legislation because of FERC's status as an independent regulatory agency.

Other measures that do not interfere with agencies' jurisdiction could be considered, such as memoranda of understanding and other commitments to complete project reviews in a timely way. A standard form or protocol could be developed to ensure that cooperative understandings are in place without compromising any agency's authority.

Innovative siting practices

Not surprisingly, most applicants prefer to use siting practices that have worked before. They believe this approach improves their chances of success, and that new approaches are risky. One reason for their caution is that mounting a transmission siting effort can be expensive, particularly if it is unsuccessful.[47] Despite this bias, innovative approaches that invest in early and more open planning and

---

[47]See the section "Description of the Transmission Siting Process," on page E-3, for details on a transmission project that Florida Power abandoned after more than a decade of effort and expenditures of $23 million.

consider a more comprehensive range of alternatives may produce better outcomes. DOE should consider funding demonstration programs in this area.

Increasing transmission capacity of existing facilities

It is increasingly well understood that for some types of transmission system needs, adding generation resources in the load center can increase transfer capacity. In addition, new technologies such as static var compensators can give operators more control over grid flows and lead to a reduction in the amount of capacity that must be reserved for "N-1" contingencies.[48] DOE could focus resources on demonstrating technological options that are available but not in common practice, such as FACTS, high-voltage direct current (HVDC), and high-temperature superconductivity (HTS), which would increase the transfer capacity of existing facilities.[49]

Identifying "best practices" for reviewing agencies

DOE could work with appropriate state-based organizations[50] to identify "best practices" for consideration by transmission siting authorities. The topics to be addressed could include:

• Open planning;

---

[48] An N-1 Contingency refers to the practice of assuring that the transmission system can withstand the change in power flows resulting from the sudden loss of any element on the system.

[49] FACTS devices are sophisticated solid-state electronic switches that allow operators to control flow on certain power lines. HVDC lines do not operate synchronously with the AC grid but can move large amounts of power over great distances with almost no losses. HTS can also move large amounts of power with almost no losses; this technology is under development. See Issue Paper Advanced Transmission Technologies by J. Hauer, T. Overbye, J. Dagle, and S. Widergren.

- Treatment of alternatives;

- Criteria for project approval, including determination of need;

- Maximum time limits;

- Strategic use of undergrounding;

- Innovative easement agreements;

- Use of mitigating measures;

- Estimating probable cost/benefit implications for affected jurisdictions; and

- Development of model rules and decision criteria.

The Tennessee Valley Authority and the federal power marketing administrations with active transmission siting responsibilities could also participate in this project and adopt the resulting practices.

Guidelines for applicants

The federal government has a great capacity to provide leadership as can be seen in many energy-related areas. For example, the Federal Energy Management Program of DOE is working to make federal buildings energy efficient, not only as good management practices for those buildings, but also to set an example. Regarding transmission siting, DOE could work with state agencies[51] and industry organiza-

---

[50]Participation by organizations such as the National Governors' Association, the Western Governors' Association, and the National Association of Regulatory Utility Commissioners would be important to the success of such a project.
[51]See previous footnote.

tions[52] to develop guidelines that would aid applicants in securing timely approval for proposed new transmission or grid-related projects. This project to develop guidelines would consider much the same subject matter as the preceding one focused on "best practices" but from the applicant's perspective. The Tennessee Valley Authority and the federal power marketing administrations with active transmission siting responsibilities could also contribute to the success of this project.

Innovative regulatory methods
Investor-owned utilities' high-voltage transmission systems are under FERC's rate-making jurisdiction.[53] Many utilities believe that rate-making incentives to build new transmission facilities are not adequate and have proposed increasing the return on investment allowed in transmission rates. There is also concern that transmission pricing should better reflect system economics and power flows. Addressing these proposals in detail is outside the scope of this paper,[54] but some comments about alternative approaches are relevant in the context of improving siting processes.

Utilities' incentives are clearly driven by the regulations that define their revenue stream. Volumetric transmission rates promote increased volume on the grid, and utilities respond in a logical way by increasing throughput on their

---

[52]Organizations such as the Edison Electric Institute, the American Public Power Association, the National Rural Electric Cooperative Association, and the Electric Power Supply Association could provide valuable assistance in the design and implementation of such a project.

[53]This is true everywhere in the contiguous United States except Texas.

[54]See the Issue Paper Alternative Business0 Models for Transmission Investment and Operation by S. Oren, G. Gross, and F. Alvardo addresses the return on equity issue. Generally, performance-based rate making for transmission service offers the prospect of improving utility incentives by bringing them into better alignment with the public interest.

systems. In some cases, congestion or reliability problems ensue, leading to calls for additional capacity. An alternative approach would be to compensate utilities fairly (at whatever rate of return on equity regulators choose) for the use of their facilities regardless of throughput. Each utility would have its transmission rates set to recover its costs plus the return and would be subject to periodic rate adjustments to true up any divergence between expected revenue and actual results. Performance incentives for reliability and service could be incorporated into the system.

Under this regulatory alternative, a transmission-owning utility has no undue bias toward growth in assets. Investments that may promote more efficient use of existing facilities and avoid the need for new facilities may be more vigorously pursued, which may align corporate incentives more closely with the public interest. FERC could actively invite utilities to experiment with this form of regulation for a defined period of time. DOE could work with FERC to develop the plan.

Another area where FERC activities could be very helpful to transmission siting is in RTO development. An RTO can become an unbiased source of accurate, publicly tested regional planning information that can help siting authorities evaluate and validate the need for a variety of grid-related investments. An RTO can also provide insight about the appropriate allocation of the costs of interstate projects and about how transmission services should be priced in order to provide accurate economic signals for grid-related investments.

SUMMARY AND CONCLUSIONS

Siting electric transmission lines is currently a state responsibility.[55] Each state has the option to address transmission siting in its own laws, and most have done so. In most states, applicants must demonstrate that proposed facilities are needed, and a state siting authority must confirm that construction of the facilities would serve the public interest. If a facility would cross state lines, approval is needed from each state affected. Additional approvals are required from federal agencies if the line would cross federally owned or controlled lands, and consent from Native American tribes is needed to cross tribal lands. The public process for reviewing and approving the siting of proposed transmission facilities is unavoidably difficult and complex because it entails fitting long-lived and highly visible structures into physical surroundings where land is already in use for other purposes. This is especially true for transmission projects that are large in geographic scale because they tend to require approvals from many affected jurisdictions.

During the past decade, most small-scale, intrastate transmission proposals have been approved without major delay or controversy. Delay and controversy have been more common in larger, interstate projects; however, approval has been obtained eventually in most cases if the applicant has been persistent and presented alternative proposals. Some parties believe that this record is misleading, and suggest that some or even many applicants have refrained from proposing large-scale, multistate transmission projects. It is difficult to verify the extent of such withholding, but there has been a striking disparity during the past decade between the level of new investment in generation and the level of new investment in transmission. This disparity suggests that some major transmission projects may indeed have been

---

[55]With the exception of the federal power marketing administrations and the Tennessee Valley Authority, which have their own siting authorities.

withheld and may not be just the result of excess capacity built in prior decades (though siting authorities should guard against the prospect of accelerated construction producing a new generation of stranded utility costs).

There are several possible reasons for withholding of proposals:

- Regional-scale transmission planning has lagged behind the development of regional-scale bulk power markets. It may be that the economic feasibility of some multistate projects is only now becoming apparent. The penalties to companies or investors who misjudge the economics of such projects can be severe.

- The transmission sector of the industry is in the midst of a fundamental reorganization. Many companies have not known whether they will remain in the transmission business or what the rules will be that will determine the rate of return on new transmission investments. It is reasonable to assume that some companies will not present new proposals until these uncertainties are resolved.

- The present state-based transmission siting process is difficult at best, particularly for large-scale projects.

Given these considerations, it is understandable that there is disagreement between those who think that the existing siting regime is basically sound but needs improvement, and those who believe that fundamental reforms are needed.

Problem Areas in the Existing Regime

Approval of a proposed transmission project is the cul-

mination of a long and complex process that can go awry for
many reasons. In addition, the transition to regional bulk
power markets may raise significant new difficulties related
to transmission siting. Some of the principal problem areas
are:

Need for regional-scale transmission planning

Although some regional plans have been developed,
many areas of the nation do not have regional plans, and
some of the plans that have been prepared are very incomplete (see the Issue Paper *Transmission Planning and the Need
for New Capacity* by E. Hirst and B. Kirby). There is an urgent
need for regional transmission plans that after public review
will confirm to prospective applicants and reviewing agencies that specific regional transmission needs have been
identified and ranked according to priority. Regional transmission planning is one of several critical functions that regional transmission organizations (RTOs) would perform, as
envisioned by FERC.

Possible need for interim transmission plans

Rather than wait for RTOs to be formed and regional
transmission plans to be developed by them, as an interim
measure it might be useful for DOE and FERC to identify
key bottlenecks and for the FERC to task administrative law
judges to work with appropriate parties in the bottleneck
areas to develop interim transmission plans. A possible benefit of such plans is that they would probably flag some
important issues affecting groups of states, and thus help to
spur the formation of cooperative regional institutions.

Need for transparent planning and systematic consideration of alter-
natives by applicants

To win approval, a transmission proposal should be

developed through a process open to participation by all interested parties and with systematic attention to a broad range of alternatives.

Need for coordination, consistency, and timeliness of federal agency reviews

Applicants and other parties cite four kinds of problems with federal agency reviews of transmission siting proposals:

1. Local or regional officials within an agency are sometimes inconsistent in their reviews of transmission projects.

2. If two or more federal agencies are reviewing a project, communication and coordination between/among them are sometimes inadequate.

3. Review of transmission proposals is sometimes given little priority in comparison to the primary mission of the agency.

4. Federal agencies sometimes wait to conduct their reviews until state reviews are completed and a final route has been proposed. This introduces the risk that a federal agency may require a route change, leading to another time- and cost-consuming iteration in the state-level process.

Need for coordination and development of a common review process

All state agencies with review responsibilities, relevant federal agencies, and tribal authorities within a region should use a common review process and coordinate reviews of transmission siting proposals. Inadequate coordination and cooperation among reviewing agencies (and the applicant) can significantly hinder the siting process and may lead to rejection of a project by one or more agencies.[56]

Need to regulate the time allowed for reviews

Many corporate parties to the transmission siting process assert that the unpredictable timing of typical state-based siting processes contributes significantly to the uncertainty hindering key business decisions in the transmission sector today. Many parties favor state and/or federal legislation setting fixed time limits (e.g., 12–18 months) for reviews. Projects not acted upon within the time period would be approved by default. The success of this approach would depend to a significant extent on the filing of a complete application at the outset, and affected agencies would probably enforce "completeness" very strictly.

Potential disagreement between states over definition of "need"

One state's definition of "need" for new transmission capacity may include transmission to enable additional electricity commerce; a neighboring state may limit "need" to transmission needed to maintain reliability.

Potential disagreement between states over whether a particular facility is needed

Even if two states have identical definitions of need, they may still not agree that a proposed facility is the best alternative for meeting a specific requirement.

---

[56]Examples include AEP's 765-kV line in Virginia and West Virginia, and the Cross Sound Connector project between Long Island, New York, and Connecticut, both of which are described above in the section "Assessment of Current Siting Regime."

Potential disagreement between states
over distribution of costs and benefits

An interstate project may fail to win all required approvals unless the affected states come to agreement about the distribution of the facility's costs and benefits. A key element of disagreement may be the time horizon over which benefits and costs are assessed.

Need for regional institutions to facilitate
the siting process for interstate projects

The western states have had extensive experience with siting interstate transmission projects, and an institutional framework is evolving under the auspices of the Western Governors' Association[57] to aid the states in dealing with such projects. In the eastern U.S., however, interstate projects have been less frequent, and, for the most part, comparable institutional frameworks remain to be developed.

Options for Improving the Transmission Siting Process

The recent debate over whether to make a federal agency, most likely FERC, responsible to some degree for siting major new transmission facilities has been healthy and useful though sometimes acrimonious. It has put all parties on notice that this process must work—it must lead to a timely determination by appropriate government agencies regarding whether proposed facilities are needed and to the approval of routes or sites for needed facilities. The debate

---

[57]The Western Interstate Energy Board, which is the energy arm of the Western Governors' Association, and the Western Conference of Public Service Commissions acted jointly in 1984 to create the Committee on Regional Electric Power Cooperation (CREPC). CREPC has representation from the regulatory commissions, energy agencies, and facility-siting agencies in the 11 states and two Canadian provinces in the Western Interconnection. Through CREPC, the western states have begun negotiations to establish a common interstate transmission-siting protocol.

has also provided impetus for a searching examination of options for improving the process. Many of these options are listed below.

Options for individual states

1. Promote or require an open, transparent transmission planning process.

2. Require project applications to address a broad range of alternatives.

3. Review and if appropriate clarify or update criteria for approval; consider whether the requirements of commerce should be recognized explicitly in determining "need" for transmission capacity.

4. If necessary, modify state law to enable siting authorities to take account of out-of-state benefits when assessing the merits of a transmission siting proposal.

5. Adopt a "one-stop" siting process. Local and county governments could use zoning to direct utility facilities to preferred locations, but they would lose the ability to reject a project. State reviews would be consolidated in the siting authority.

6. Set a maximum time limit (e.g., 12 or 18 months) for reviews by state or local agencies.

7. State clearly what materials must be included in an application, and refuse to initiate a review until an application is complete.

8. Promote use by applicants of both deterministic and probabilistic planning methods.

9. Promote more consistent use of "rolled-in"" and "cost causation" approaches to recovering the cost of new grid-related investments, to minimize either favoring or disadvantaging particular technological alternatives.

10. Promote innovative approaches to meeting transmission grid needs.

11. Emphasize to prospective applicants that undue minimization of transmission project costs can be self-defeating.

Regional options

All of the state-level options listed above have regional significance; that is, if they were considered and applied by all states in a given region, the result would probably be greater regional consistency and efficacy in siting policies and practices. The options below focus on development of regional institutions that could, among other objectives, promote such consistency and efficacy. States, federal land management agencies, and Native American tribes should consider the following options:

1. Support and participate in open, transparent regional transmission planning.

2. Promote the development of cooperative regional transmission siting institutions that would have two key missions:
    (a) Develop elements of a common siting process, usable by most and if possible all reviewing agencies; and
    (b) Maintain parallel processes among reviewing agencies, utilizing consistent information, identifying in-

formation gaps or possible points of disagreement early, and ensuring that these are addressed by a scheduled calendar date.

3. Agree that if an agency fails to complete its review by a scheduled calendar date, the application is approved by default.

4. Consider whether a regional organization with decision-making powers should be established to address some energy regulatory matters on a regional basis (i.e., over-sight of system planning, siting and permitting, rate regulation, or other matters).

Federal options

Most of the options listed above could be aided through specific federal actions, including:

1. Establish broad federal support for open, transparent regional-scale planning to address generation require-ments, generation siting considerations, transmission requirements, and related issues.

2. As an interim measure while waiting for RTOs to be formed and regional transmission plans to be prepared by them, DOE and FERC could act jointly to identify key transmission bottlenecks, and FERC could task ad-ministrative law judges to work with appropriate par-ties in each bottleneck area to prepare an interim transmission plan by a specific date.

3. Improve the process for the review of transmission sit-ing proposals by federal land management agencies. Several sub-options could be implemented by a Presi-dential executive order:

(a) Direct federal land managers and other relevant agencies to support and participate in common and coordinated state or regional processes for timely review of proposals for new transmission facilities requiring federal approval.

(b) Require all federal reviews to be completed within 18 months after the filing of a complete application. Applications not acted upon within 18 months would be approved by default.

(c) Establish training programs on the national significance of the transmission grids and related issues, and make these programs mandatory for federal officials authorized to approve or reject transmission siting proposals.

(d) Create special staff groups in the headquarters of appropriate federal agencies to work jointly to prepare consolidated, multi-agency reviews of proposed transmission projects.

(e) Direct that if two or more agencies have jurisdiction over a proposed transmission project, the Office of Management and Budget shall designate one of them as the lead agency, responsible for coordinating the preparation of a timely joint review of the proposal. (Note: An alternative to this arrangement would be to enact federal legislation making FERC responsible for coordination of all federal reviews of transmission projects, as described below.)

4. Seek federal legislation that would:
   (a) Direct the Secretary (DOE) or FERC to initiate a rulemaking to establish criteria for the identification of transmission bottlenecks (or projects to ease such bottlenecks) of national or regional importance.

(b) Affirm that for projects designated to be of national or regional importance, an applicant would have the right to petition FERC to assume a backstop role in the event that a state or tribal reviewing agency does not act to approve or deny the project within 18 months after the filing of a complete application. (A stronger but more controversial and less predictable formulation would be to empower applicants to petition FERC when a state, tribal, or federal reviewing agency acts in the allotted time but rejects the application. "Forum shopping" could become a significant problem if applicants could always turn to FERC for a second opinion. If this version were adopted, items c and d below would have to be modified for consistency.)

(c) Empower FERC to decline a petition for cause, and limit FERC's role to serving as a backstop for the agency that has not acted, without affecting the actions or responsibilities of other reviewing agencies.

(d) Direct that FERC shall be the lead agency for coordinating all reviews of proposed transmission facilities by federal agencies, that other affected federal agencies shall participate as cooperating agencies, and that the cooperating agencies will retain their existing authorities with respect to the issuance of permits for lines crossing lands under their jurisdiction.

5. Undertake a DOE project, jointly with NGA, WGA, NARUC, and other appropriate state-based organizations to articulate a set of "best practices" related to transmission siting for consideration by all states.

6. Undertake a DOE project, jointly with appropriate state agency organizations and industry trade associations, to articulate a set of guidelines for applicants, designed to increase the likelihood of approval of proposed new transmission or grid-related projects.

7. Undertake a DOE demonstration program to support applicants in taking innovative approaches to transmission siting proposals (e.g., treatment of alternatives, use of innovative or little-used technologies, imaginative use of mitigating measures, etc.).

8. Undertake a DOE demonstration program to support the use of new or under-used methods and technologies for increasing the transmission capacity of existing facilities.

9. Support FERC efforts to improve the incentives of transmission-owning companies and other potential developers of new transmission capacity or other grid-related projects through performance-based regulation.

ACKNOWLEDGMENTS

We thank Joe Eto at the Lawrence Berkeley National Laboratory for his cheerful confidence and support and the writers of the other papers in this series for their insights. We are grateful for the support of Department of Energy staff, particularly the expertise of Paul Carrier and Jimmy Glotfelty in bringing a large and complex enterprise together. We thank the many individuals who provided invaluable background information, commentary, and suggestions, including the speakers at the three public hearings the DOE convened for this project.

Particular thanks go to the following individuals who took time to talk with or write to us about how transmission siting really works in the United States: Henry Bartholomew, Martin Bettman, James Boone, Jeffrey Butler, Roy Buxton, Daniel Cearfoss, Jr., Mark Cooper, Clark Cotton, John Cowger, Mark Doljac, Mel Eckhoff, Steve Ferguson, Alan Fiksdal, Webster Gray, Brent Hare, Cheryl Harrington, Doug Hartley, Rich Hoffman, Fred Hoover, John Hynes, Blaine Keener, Ron Klinefelter, Klaus Lambeck, Doug Larson, Rick Loughery, Mark MacLeod, Earl Melton, Martha Morton, John Owens, Sandi Patti, Gary Porter, Prasad Potturi, Lee Rampey, Raj Rana, Gene Reeves, Tom Ring, Jose Rotger, Wayne Shirley, Alison Silverstein, Julie Simon, Udaivir Sirohi, Karen Skinner, Jeff Steir, Don Stursma, Massoud Tahamtani, Bob Timmer, John Underhill, Steve Walton, Randy Watts, Carl Weinberg, Thomas Welch, Jim Wexler, Connie White, and Kim Wissman.

References

Alvarado, F. and S. Oren. 2002. *Transmission System Operations and Interconnection. National Transmission Grid Study*, U.S. Department of Energy.

Bailey, M. and C. Eaton. 2001. "Moving Toward Seamless Energy Markets: Evidence from the Northeast." *Electricity Journal*, v. 14, #6, July.

Buxton, P.R. Engineering Department. Illinois Commerce Commission. personal communication regarding Illinois Commerce Commission Docket 92-0121.

Costello, K.W. 2001. "Interregional Coordination versus RTO Mergers: A Cost-Benefit Perspective." *Electricity Journal*, v. 14, #2, March.

Doljac, M. August 14, 2001. personal communication. Kansas Corporation Commission.

Electric Utility Week (McGraw-Hill). Oct. 2, 1995, p. 15.

Electric Utility Week (McGraw-Hill). April 1, 1996, p. 14.

Electric Utility Week (McGraw-Hill). September 17, 2001, p. 17.

Fox-Penner, P. 2001. "Easing Gridlock on the Grid: Electricity Planning and Siting Compacts," *Electricity Journal*, v14, #6, July.

Hauer J., T. Overbye, J. Dagle, S. Widergren. 2002. *Advanced Transmission Technologies*. U.S. Department of Energy.

Hirst, E. and B. Kirby. 2002. *Transmission Planning and the Need for New Capacity*. U.S. Department of Energy.

Mississippi State Code 77-3-14.

Moskovitz, D. 2001. "Distributed Resource Distribution Credit Pilot Programs: Revealing the Value to Consumers and Vendors." National Renewable Energy Laboratory, September 2001.

O'Donnell, K. 2000. "Worth the Wait, But Still at Risk." *Public Utility Fortnightly*, v. 138, #9, p. 20, May 1.

Oren, S., G. Gross, and F. Alvarado. 2002. *Alternative Business Models for Transmission Investment and Operation*. U.S. Department of Energy.

Stavros, R. 2000. "Transmission 2000: Can ISOs Iron Out the Seams?" *Public Utility Fortnightly*, v. 138, #9, p. 24, May 1.

Conceptual Plans for Electricity Transmission in the West, Report to the Western Governors Association, August 2001, p. 13.

U.S. Department of Commerce. 2000. *Statistical Abstract of the United States*. December 2000.

U.S. Department of Energy. 1998. *Maintaining Reliability in a Competitive U.S. Electricity Industry, Final Report of the Task Force on Electric System Reliability to the Secretary of Energy Advisory Board*, Washington, DC.

Advanced Transmission Technologies

(National Transmission Grid Study)

John Hauer
Pacific Northwest National Laboratory
Richland, Washington
Tom Overbye
University of Illinois at Urbana-Champaign
Urbana, Illinois
Jeff Dagle
Pacific Northwest National Laboratory
Steve Widergren
Pacific Northwest National Laboratory

INTRODUCTION

This paper discusses the use of advanced technologies to enhance performance of the national transmission grid (NTG). We address present and developing technologies that have great potential for improving specific aspects of NTG performance, strategic impediments to the practical use of these technologies, and ways to overcome these impediments in the near term.

Research and development (R&D) infrastructure serving power transmission is as badly stressed as the grid itself, for many of the same reasons. The needs are immediate, and the immediate alternatives are few. Timely and strategically effective technology reinforcements to the NTG need direct, proactive federal involvement to catalyze planning and execution. Longer-term adjustments to the R&D infrastructure may also be needed, in part energy policy can evolve as the NTG evolves.

Technology and a coordinated national effort are only two of the elements necessary for timely resolution of the problems facing the national energy system. Sustainable solutions require careful balancing between generation and transmission, profit and risk, the roles of public and private institutions, and market forces and the public interest. There is a vast body of information and opinion on these issues. A recent white paper by EPRI (formerly known as the Electric Power Research Institute) clearly lays out the broad issues and a comprehensive inventory of technology options for enhancing the grid, including detailed assessments of their direct costs and benefits. Titled "The Western States Power Crisis: Imperatives and Opportunities," (EPRI 2001), this document notes that "...the present power crisis—most evident in the Western states but potentially a national problem—requires a fundamental reassessment of the critical interactive role of technology and policy in both infrastructure and markets" (EPRI 2001). Similar assessments of needs and solutions, many of which arrive at similar conclusions, are found in a series of studies extending back to 1980 (DOE 1980). A widely shared view concerning the urgency of technology solutions is provided in Scherer 1999.

The strategic need is not just for new technology in the laboratory but for an infusion of improved, cost-effective technology to work in the power system. The chief impediments to infusion are institutional and can be resolved by a proactive national consensus regarding institutional roles. Until this consensus is achieved, the lack of cohesion between technology and policy may be disruptive for continued development of the NTG and the infrastructures that it serves.

This issue paper discusses the use of new technologies to enhance the performance of the NTG, as follows:

- Background on power system operation in general and the specifics of the NTG.

- The new demands being placed on the NTG and outlines the technology needed to address these demands.

- The impact of existing institutional frameworks on the application of new technology to the transmission grid.

- The strategic challenges that can be addressed through accelerated use of selected new technologies.

- The institutional issues associated with moving new technology from the research laboratory to deployment in the grid.

- A summary of some of the options discussed in the paper.

- Appendix A is an extensive (though not exhaustive) list of new technologies that could be applied to the NTG.

BACKGROUND

The transition to open electrical energy markets is stressing the NTG beyond its design capabilities. Less conspicuously, this transition is also stressing the management infrastructure by which transmission facilities are planned, developed, and operated. Stresses on this infrastructure are a major strategic impediment to the focused development and timely deployment of technical solutions to shortfalls in national grid capacity. The subsections below give some transmission system background that is necessary to understand these technological issues.

Power System Components and Reciprocal Impacts

The power system has three components: generation, load, and transmission. Electric power is produced by generators, consumed by loads, and transmitted from generators to loads by the transmission system. Typically, the "transmission system" (or "the grid") refers to the high-voltage, networked system of transmission lines and transformers. The lower-voltage, radial lines and transformers that actually serve load are referred to as the "distribution system." The voltage difference between the transmission system and the distribution system varies from utility to utility; 100 kV is a typical value. This paper focuses only on advanced technologies for the transmission system.

It is important to understand reciprocal impacts among the transmission system, load, and generation. Because the transmission system's job is to move electric power from generation to load, any technologies that change or redistribute generation and/or load will have a direct impact on the transmission system. This can be illustrated using a simple two-bus, two-generator example shown in one-line form in Figure 1. The solid lines represent the buses, the circles represent the generators, and the large arrow represents the aggregate load at bus 2. Three transmission lines join the generator at bus 1 to the load and generation at bus 2. Superimposed on the transmission lines are arrows whose sizes are proportional to the flow of power on the lines. The pie charts for each line indicate the relation between the loading on each line and its rated capacity. The upper and middle transmission lines have a rating of 150 MVA, and the lower line has a rating of 200 MVA. In addition, we assume that the bus 1 generation is more economical than the generation at bus 2, and the entire load is being supplied remotely from the bus 1 generator. With a bus 2 load of 420 MW, the power distributes among the three lines based on their impedances (which are not identical), so the upper line is loaded at 67 percent, the middle at 89 percent, and the lower at 100 percent. Note: there are 13 MW of transmission line losses in this case.

Transfer Capacity

A natural question to ask is: what is the transfer capacity of the transmission system described in Figure 1? That is, how much power can be transferred from bus 1 to bus 2? The answer is far from straightforward. At first glance, the transfer capacity appears to be 420 MW because this amount of power causes the first line to reach its limit. However, this answer is based on the assumption that all lines are in service. As defined by the North American Electric Reliability Council (NERC), transfer capacity includes consideration of reliability. A typical reliability criterion is that a system be able to withstand the unexpected outage of any single system element; this is known as the first contingency total

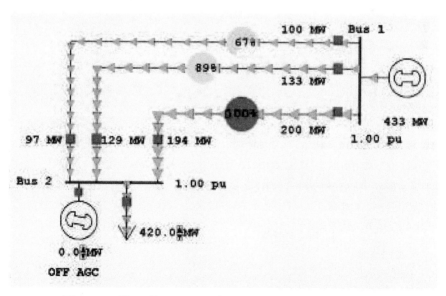

Figure 1: Two-Bus Example with No Local Generation

transfer capability (FCTTC). Based on this criterion, Figure 2 shows the limiting case with an assumed contingency on the lower line, which results in a transfer capability of only 252 MW. Which number is correct?

The answer depends on system operational philosophy and on the availability of high-speed system controls. If the operational philosophy requires that no load be involuntarily lost following any individual contingency, and if there are no mechanisms to quickly increase the generation at bus 2, voluntarily decrease the load at bus 2, or redistribute the flow between the remaining upper two lines, then the limit would be 252 MW. With these limitations, the only way to increase the transfer capacity would be to construct new lines.

However, if we relax one or more of these conditions, the transfer capacity could be increased without construction of new lines. For example, one approach would be to provide at least some of the bus 2 load with incentives so that,

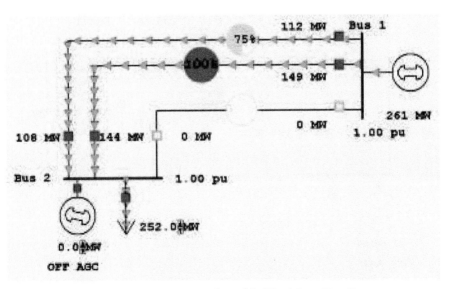

Figure 2: Two-Bus Example with Limiting Contingency

following the contingency, some customers on bus 2 would voluntarily curtail their loads. Incentives might involve price-feedback mechanisms or agreements to allow the system operator to curtail load through some type of direct-control load management or interruptible demand. Another approach would be to have a mechanism for quickly committing some local bus 2 generation. Availability of local generation reduces the net loading on the transmission system and can increase its capacity. A third approach would be to use advanced power electronics controls such as flexible AC transmission system (FACTS) devices to balance the load between the upper two lines.

The unifying themes of these alternative approaches are knowledge about the real-time operation of the system, availability of effective controls, and an information infrastructure that permits effective use of the controls. To understand these themes, it is important to understand the complexity of the actual national transmission grid.

Complexity of the National Transmission Grid

The term "national transmission grid" is something of a misnomer. The North American transmission grid actually consists of four large grids, each primarily a synchronous alternating current (AC) system. Together, these four grids span parts of three sovereign countries (U.S., Canada, and Mexico). By far the largest grid is the Eastern Interconnection, which supplies power to most of the U.S. east of the Rocky Mountains as well as to all the Canadian provinces except British Columbia, Alberta, and Quebec. The Western Interconnection supplies most of the U.S. west of the Rockies, as well as British Columbia, Alberta, and a portion of Baja, California. The remaining two grids are the Electric Reliability Council of Texas (ERCOT), which covers most of Texas, and the province of Quebec. In contrast to the two-bus example presented above, the Eastern and the Western Interconnections contain tens of thousands of high-voltage buses and many thousands of individual generators and loads. Because the individual grids are asynchronous with one another, no power can be transferred among them except in small amounts through a few back-to-back direct current (DC) links. Several major DC transmission lines are also used within the individual grids for long-distance power transfer.

At any given time the loading on the grid depends on where power is being generated and consumed. Load is controlled by millions of individual customers, so it varies continuously. Because electricity cannot be readily stored, generation must also vary continuously to track load changes. In addition, the impedances of the many thousands of individual transmission lines and transformers dictate grid loading. With several notable exceptions, there is no way to directly control this flow—electrons flow as dictated by the laws of physics. Because electricity propagates through the network very rapidly, power can be transferred almost instantaneously (within seconds) from one end of the grid to the other. In general, this interconnectivity makes grid operations robust and reliable. However, it also has a detrimental effect if the grid fails; failures in one location can quickly affect the entire system in complex and dramatic ways, and large-scale blackouts may result.

The grid's ability to transfer power is restricted by thermal flow limits on individual transmission lines and transformers; minimum and maximum limits on acceptable bus-voltage magnitudes; and region-wide transient, oscillatory, and voltage-stability limitations. Given NERC's reliability requirements, these limits must be considered not only for current and actual system operating point but also for a large number of statistically likely contingent conditions as well. The complexity of maximizing the power transfer capability of the grid while avoiding stressing it to the point of collapse cannot be overstated.

Technologies to Increase Transfer Capacity

The goal of this issue paper is to examine technologies that can be used to increase the grid's power-transfer capability. This increase can be achieved by a combination of direct technical reinforcements to the grid itself along with indirect information and control reinforcements that improve grid management practices and infrastructure.

Direct reinforcement of the grid includes new construction and broad use of improved hardware technology. Strategic decisions regarding these two types of improvements are a function of grid management—planning, development, and operation. Grid management involves recognizing transmission needs, assessing options for meeting those needs, and balancing new transmission assets and new operating methods. Timely development and deployment of requisite technology are essential to reinforcing the grid. Requisite technology may not mean new technology. There is a massive backlog of prototype technology that can, given means and incentives, be adapted to power system applications.

Indirect grid reinforcement includes improving grid management by means of technology. Historically, the transmission system was operated with very little real-time information about its state. During the past few decades, advances in computer and communication technology in general and SCADA (supervisory control and data acquisition) and EMS (energy management system) technology in particular have greatly improved data capabilities. Significant real-time data are now available in almost every control center, and many centers can conduct advanced on-line grid analysis. Despite these improvements, more can and should be done. In the control center, additional data need to be collected, better algorithms need to be developed for determining system operational limits, and better visualization methods are needed to present this information to operators. Beyond the control center, additional system information needs to be presented to all market participants so that they can make better-informed decisions about generation, load, and transmission system investments.

Institutional Issues that Affect Technology Deployment In order to effectively discuss the role of advanced transmission technologies, we have to consider how their deployment is either hindered or encouraged by institutional issues. Ultimately, the bottom line is economics —technologies that are viewed as cost effective will be used, and those that are considered too costly will not. The issue of cost is not simple; public policy must address how costs and benefits should be allocated. For example, it is difficult to beat the economics of traditional overhead transmission lines for bulk power transfer. The lines are cheap to build and entail relatively few ongoing expenses. But the siting of new transmission lines is not so simple; right of way may be difficult to obtain, and new lines may face significant public opposition for a variety of reasons from aesthetic to environmental (for a detailed discussion, see Issue Paper *Transmission Siting and Permitting* by D. Mayer and R. Sedano.) Advanced technologies can reinforce the grid, minimizing the need for new overhead lines, but usually at higher cost than would be paid to build overhead lines. The challenge is to provide incentives that will encourage the desired transmission investments.

Unfortunately, in recent years the uncertainties associated with electricity industry restructuring have hampered progress in transmission reinforcement. The boundaries between responsibilities for operation and planning were once clearly delineated, but these responsibilities are now shifting to restructured or entirely new transmission organizations. This process is far from complete and has greatly weakened the essential dialogue between technology developers and users. Development of new technology must be closely linked to its actual deployment for operational use. Together, these activities should reflect, serve, and keep pace with the evolving infrastructure needs of transmission organizations. The current uncertainty discourages this cohesiveness.

The details and the needs of the evolving infrastructure for grid management are unclear, and all parties are understandably averse to investments that may not be promptly and directly beneficial. Some utilities are concerned that transmission investments may be of greater benefit to their competitors than to themselves. In the near term, relief of congestion may actually harm their businesses. As a result of such forces, many promising technologies are stranded at various points in route from concept to practical use. Included are large-scale devices for routing power flow on the grid, advanced information systems to observe and assess grid behavior, real-time operating tools for enhanced management of grid assets, and new system planning methods that are robust in relation to the many uncertainties that are present or are emerging in the new power system.

Another important issue is that some technologies that would enable healthy and reliable energy commerce are not perceived as profitable enough to attract the interest of commercial developers. Special means are needed to develop and deploy these technologies for the public good. Involvement by the federal utilities and national laboratories may be necessary for timely progress in this area, as well as a broadening of some activities of EPRI or similar umbrella organizations focused on energy R&D along with development of better mechanisms to spur entrepreneurial innovation.

NEW DEMANDS ON THE TRANSMISSION GRID

The core objective underlying electricity industry restructuring is to provide consumers with a richer menu of potential energy providers while maintaining reliable delivery. Restructuring envisions the transmission grid as flexible, reliable, and open to all exchanges no matter where the suppliers and consumers of energy are located.

However, neither the existing transmission grid nor its current management infrastructure can fully support such diverse and open exchange. Transactions that are highly desirable from a market standpoint may be quite different from the transactions for which the transmission grid was designed and may stress the limits of safe operation. The risks they pose may not be recognized in time to avert major system emergencies, and, when emergencies occur, they may be of unexpected types that are difficult to manage without loss of customer load.

The transmission system was originally constructed to meet the needs of vertically integrated utilities, moving power from a local utility's generation to its customers. Interconnections between utilities were primarily to reduce operating costs and enhance reliability. That is, if a utility unexpectedly lost a generator, it could temporarily rely on its neighboring utilities, reducing the costs associated with having sufficient reserve generation readily available. The grid was not designed to accommodate large, long-distance transfers of electric power.

One of the key problems in managing long-distance power transfers is an effect known as "loop flow." Loop flow arises because of the transmission system's uncontrollable nature. As power moves from seller to buyer, it does not follow any prearranged "contract path." Rather, power spreads (or loops) throughout the network. As an example,

Figure 3 shows how a transmission of power from a utility in Wisconsin to the Tennessee Valley Authority (TVA) would affect lines through a large portion of the Eastern Interconnection. A color contour shows the percentage of the transfer that would flow on each line; lines carrying at least two percent of the transfer are contoured. As this figure makes clear, a single transaction can significantly impact the flows on hundreds of different lines.

The problem with loop flow is that, as hundreds or thousands of simultaneous transactions are imposed upon the transmission system, mutual interference develops, producing congestion. Mitigating congestion is technically difficult, and very complex problems emerge when paths are long enough to span several regions that have not had to coordinate such operations in the past. These problems include (but are not limited to) the lack of: effective procedures, operating experience, computer models, and integrated data resources. The sheer volume of data and information concerning system conditions, transactions, and events is overwhelming the existing grid management's technology infrastructure.

Increasing the transfer capacity of the NTG will require combined application of hardware and information technologies. On the hardware side, many technologies can be developed, refined, or simply installed to directly reinforce current transmission capabilities. These technologies range from passive reinforcements (such as new AC lines built on new rights of way or better use of existing AC rights of way by means of innovative device configurations and materials) to super-conducting equipment to large-scale devices for routing grid power flow. High-voltage direct current (HVDC) and FACTS technologies appear especially attractive for flow control. Effectively deployed and operated, such technologies can be of great value in extending grid

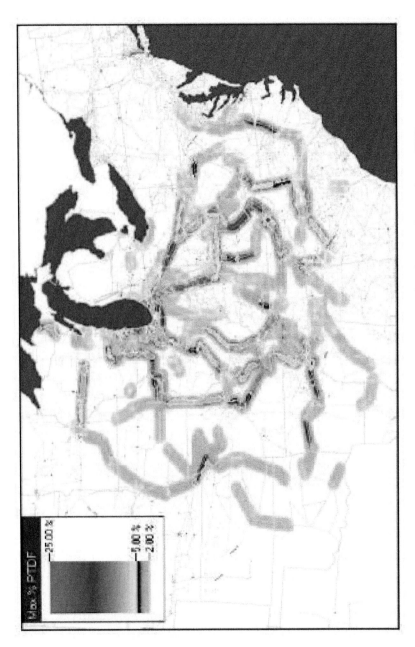

Figure 3: Loop Flow of Power Transfer from Wisconsin to TVA

capabilities and minimizing the need for construction of new transmission.

The strategic imperative, however, is to develop better information resources for all aspects of grid management— planning, development, and operation. Technologies such as large-scale FACTS generally require the support of a wide-area measurement system (WAMS), which currently exists only as a prototype. Without a WAMS, a FACTS or any major control system technology cannot be adjusted to deliver its full value and, in extreme cases, may interact adversely with other equipment. FACTS technology can provide transmission "muscle" but not necessarily the "intelligence" for applying it.

An example of the information that a WAMS can provide is shown in Figure 4. Review of data collected on the Bonneville Power Administration (BPA) WAMS system following a grid disturbance on August 10, 1996, suggests that the information that system behavior was abnormal and that the power system was unusually vulnerable was buried within the measurements streaming into and stored at the control center. Had better tools been available at the time, this information might have given system operators approximately six minutes' warning of the event that triggered the system breakup (PNNL 1999).

Better information is key to better grid management decisions. The next subsection addresses the kinds of information gaps in current grid management.

Information Gaps in Grid Management

As the grid is operated closer to safe limits, knowing exactly where those limits are and how much operating margin remains becomes increasingly important. Both limits and margins must be estimated through computer modeling and combined with operating experience that the models

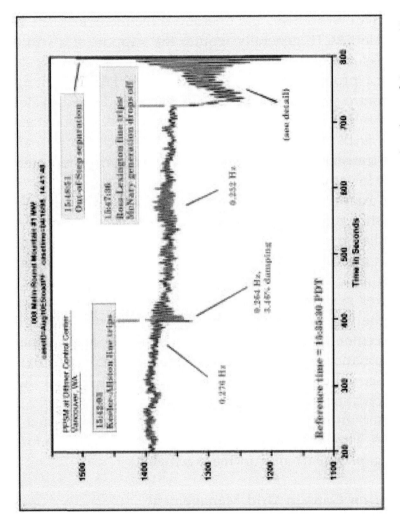

Figure 4. Possible warning signs of the Western Systems breakup of August 10, 1996 – an example of information available from WAMS

might not and often cannot reflect.

The "edge" of safe operation is defined by numerous aspects of system behavior and is strongly dependent on system operating conditions. Some of these conditions are not well known to system operators, and even those that are known may change abruptly. Important conditions include network loading, operating status and behavior of critical transmission elements, behavior of electrical loads, operating status and behavior of major control systems, and interactions between the grid and the generators connected to it. Full performance of the transmission grid requires that generators provide adequate voltage support plus a variety of dynamic support functions that maintain power quality during normal conditions and assist the system during disturbances.

All of these conditions have become more difficult to anticipate, model, and measure directly. Industry restructuring has exacerbated these difficulties by requiring that transmission facilities be managed with a minimum of information concerning generation assets. To borrow a phrase from EPRI (2001), this is one of many areas where there is a "critical interactive role" between "technology and policy." Many cases in recent years have revealed that the "edge" of safe grid operation is much closer than planning models had suggested. The Western System breakups of 1996 are especially notable in this respect (see Figure 5), but there have been less conspicuous warnings before and since (PNNL 1999). Uncertainties regarding actual system capability is a known problem of long-standing, and it has counterparts throughout the NTG.

Developing and maintaining realistic models for power system behavior is technically and institutionally difficult, and it requires higher-level planning technology than has previously been available. An infusion of enhanced planning

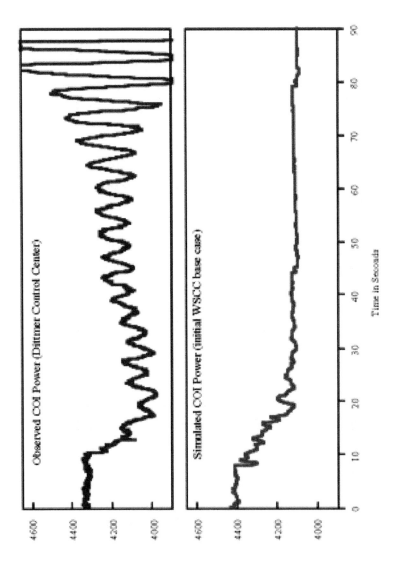

Figure 5. Modeling failure for Western System breakup of August 10, 1996. (MW on California-Oregon Interconnection)

technology—plus knowledgeable staff to mentor its development and use—is necessary to support timely, appropriate, and cost effective responses to system needs. Better planning resources are the key to better operation of existing facilities, to timely anticipation of system problems, and to full realization of the value offered by technology enhancements at all levels of the power system.

Challenges and Opportunities in Network Control

As noted earlier, the existing AC transmission system cannot be directly controlled; electric flow spreads through the network as dictated by the impedance of the system components. For a given set of generator voltages and system loads, the power-flow pattern in an electrical network is determined by network parameters. Control of network parameters in an AC system is usually quite limited, so scheduling of generators is the primary means for adjusting power flow for best use of network capacity. When generator scheduling fails, the only alternative is load control, either through voltage reductions or suspension of service. Load control can be necessary even when some lines are not loaded to full capacity.

A preferred solution would be a higher degree of control over power flow than is currently possible, which would, permit more effective use of transmission resources. Conventional devices for power-flow control include series capacitors to reduce line impedance, phase shifters, and fixed shunt devices that are attached to the ends of a line to adjust voltages. All of these devices employ mechanical switches, which are relatively inexpensive and proven but also slow to operate and vulnerable to wear, which means that it is not desirable to operate them frequently and/or use a wide range of settings; in short, mechanically switched devices are not very flexible controllers. Nonetheless, they are still the primary means used for stepped control of high power flows.

HVDC transmission equipment offers a much greater degree of control. If the support of the surrounding AC system is sufficient, the power flowing on an HVDC line can be controlled accurately and rapidly by means of signals applied to the converter equipment that changes AC power to DC and then back to AC. In special conditions, HVDC control may also be used to modify AC voltages at one or more converters. This flexibility derives from the use of solid-state electronic switches, which are usually thyristors or gate turn-off (GTO) devices.

Although HVDC control can influence overall power flow, it can rarely provide full control of the power flowing on particular AC transmission lines. However, conventional power-flow controllers that are upgraded to use electronic rather than mechanical switches can achieve this control. This upgrade opens the way to a broad and growing class of new controller technology known as FACTS. Many engineers regard HVDC technology as a subset of FACTS technology.

Increasingly, load itself is becoming a fast-acting transmission control device. Some degree of load control has been available for decades through interruptible rates, time-of-day rates, and demand-side management programs. A new possibility is use of real-time price feedback to loads in order to rapidly tailor the flow of power on the transmission grid, perhaps encouraging demand in one location while inhibiting it elsewhere. Advances in communication that can rapidly convey changing electricity prices to industrial and commercial users facilitate this control.

Short-term energy storage can aid in power flow control. Recent work shows that even a small amount of storage can significantly enhance the performance of some FACTS devices, and past research has shown that controllable storage devices have many applications in control of power quality and system dynamics (De Steese and Dagle 1997). These applications are addressed in later sections of this paper.

It should be noted that FACTS technology is still not entirely mature even though it is based on concepts that are two decades old. As has been the case with several other promising technologies, FACTS has not been utilized by the electricity industry at the rate that its apparent technical merits would justify. A number of lessons can be drawn from this. One is that innovative technologies compete against technologies that are already in place and are better understood. Many utilities view FACTS as not cost effective because of their high installation price; traditional, passive AC devices are perceived to have a cost advantage. Furthermore, though controller-based options for grid reinforcement are attractive, they are not well understood, and operating experience with another innovative technology, HVDC systems, suggests that use of new control devices may result in significant and unforeseen interactions with other equipment. Although these problems can be largely addressed with WAMS, their costs are unclear, and the consequences of controller malfunctions can be very serious. Some legal opinion holds that the liabilities from such malfunctions will be substantially greater than those faced by utilities before industry restructuring (Fleishman 1997, Roman 1999). Such considerations weigh on the side of grid reinforcement through less technically demanding means even though the return on investment may also be smaller.

Very few utilities are in a position to break this impasse as the management functions for which high-level technologies like FACTS are of primary relevance are passing from the utilities to a newly evolving infrastructure based upon

Regional Transmission Organizations (RTOs), Independent System Operators (ISOs), and other entities. This transition is far from complete in most areas of the U.S., and as yet there is no "design template" for the nature and the technology needs of this new infrastructure.

THE EVOLVING INFRASTRUCTURE FOR TRANSMISSION MANAGEMENT

The movement of operation and planning responsibilities from their place in the vertically integrated utility structure to the evolving new and restructured organizations has greatly weakened the essential dialogue between technology developers and users. Although this paper focuses on advanced transmission technologies, these technologies cannot be adequately discussed outside the context of the institutional framework within which they will be used. Development of advanced technology must be closely linked to its actual deployment for operational use. Together, these activities should reflect, serve, and keep pace with the evolving infrastructure needs of transmission organizations. Frameworks that discourage technology deployment will eventually inhibit its development. Unfortunately, the current uncertainty has produced exactly this effect.

To simplify our discussion, we assume that primary responsibility for grid management is assigned to an RTO. The following unknowns are of special concern:

- definition of RTO functions and resources,

- relationship between RTO and control areas,

- access to and sharing of operational information, and

- timeline for deployment of the supporting infrastructure for RTO operations.

Uncertainties about the evolving institutional framework for transmission management impede timely development and deployment of requisite technology. Key unknowns include:

- what functionalities require technical support and where they will be located within the overall infrastructure;

- what level of technological sophistication can be rationalized, accommodated, and supported at specific locations within the grid management infrastructure;

- how to accommodate the risks associated with operational use of prototypes;

- what extensions or refinements may be needed before particular technologies can provide full value in power system environments; and

- what the role will be of the RTO and other grid management entities in the overall R&D infrastructure serving power transmission needs.

Resolving these uncertainties in a timely manner may require that national energy policy address the infrastructure of transmission management. For the immediate future, the best course may be for policy makers to seek counsel from entities that are still involved in higher levels of grid management.

PERFORMANCE CHALLENGES TO A NEW GENERATION OF TRANSMISSION TECHNOLOGY

Many technologies, some surprising, are applicable to large power systems. Some hardware whose application to power systems may not be obvious at first include: acoustical radar to locate buried objects, radiation sensors to detect incipient failure in connectors or insulators, robotic vehicles (including unmanned aircraft) to examine the condition of transmission lines, specialized devices to mitigate the waveform pollution associated with some lighting technologies, the NASA Advanced Composition Explorer satellite (located more than one million miles from earth) to provide early warning of geomagnetic storms, and intruder alarms at unmanned facilities. Life sciences applications include study of: the biological effects of electromagnetic fields, the environmental impacts of a proposed transmission line on forest cover and wildlife, the function of naturally occurring microbes that can safely digest toxic spills, and the social/biological factors involved in management of large river systems. We can add to this a vast array of applications in materials science, advanced hardware and fuels, information systems, mathematical modeling and analysis, process automation, risk management, and decision support systems.

This paper's purpose is not to inventory the possible technology options. Recent studies by EPRI (EPRI 1997, EPRI 2001) present massive inventories with projections of likely merits, and a long series of DOE studies examines the subject from the perspective of national needs (DOE 1980-2000). The opportunities have not changed much in a decade, but the needs have become much more acute.

The subsections below list the strategic challenges that can be addressed through enhanced technology. Each challenge is stated as a functionality that will improve the overall performance of the NTG. Candidate technologies to meet each challenge are briefly discussed, and their current state of development is noted. An extensive, partial listing of new equipment technologies that could be applied to the NTG is given in Appendix A.

Technology Challenge #1:
Broader Coordination of Grid Management

DOE's National Power Grid Study of 1980 notes that "Coordinated power system planning, development and operation results in reduction in fixed costs, reduction in operating costs, lower risks, and better utilization of natural resources." The report also lists impediments to full realization. The issues raised then have been rearticulated many times since; they are persistent, basic forces in the development of large power systems.

What has changed is the context within which these forces operate. There is now an artificial information barrier between generation and transmission, and coordination across that barrier is indirect (e.g., based upon market signals). However, direct coordination across broad geographical areas has become much more feasible from a technical standpoint and is directly consistent with the objectives of industry restructuring and the effective functioning of the national transmission grid.

A recent Federal Energy Regulatory Commission (FERC) directive assigning ultimate responsibility for grid management to a few "mega-RTOs" is a step toward the institutional framework needed for truly wide area management of the NTG. Although the details of this framework are still to be worked out, information technologies will be key to the infrastructure. Integrated computer models must quickly and accurately support power-flow calculation, risk assessment, and emergency management across broad areas of North America where such activities are now performed piecemeal. Modeling studies must be reinforced by measured information, which is also needed to assure the validity of the models. Great volumes of operational data must be integrated and sifted for indications of hidden problems or to facilitate general grid management decisions. High capacity data links are needed among control centers and RTOs. High capacity information links of a different kind are needed to achieve "virtual work team" collaboration among supporting staff who may be located at widely separated locations and institutions. All of these improvements must be made with close attention to the overall security of the information process and facilities.

One approach to real-time operational system data would be to continue the current utility strategy of treating practically all such data as proprietary. Currently, only a small group of (often overworked) utility employees has access to system operational data. Although the reasons that utilities would like to keep the details of their operations hidden from public scrutiny are clear, a significant lesson from the recent electricity crisis in California is that when the grid fails, the public pays the price. Furthermore, the shared nature of the transmission grid and the fact that problems in one area can rapidly propagate throughout the entire grid make the electricity industry unique. The data that are public by federal mandate, such as FERC Form 715 filings, are helpful, but errors, such as base-case-limit violations, restrict the usefulness of these data. The release of highly processed information, such as the posting of available transmission capacity (ATC) or Locational Marginal Price (LMP) data on the Open Access Same-Time Information System (OASIS), is also helpful, but the calculations are impossible to verify or extend if the raw source data are not available.

An alternative approach would be public posting of

near-real-time operational data. FERC did not prohibit access by generators to transmission data; rather, it required that such access be non-discriminatory. Freeing the data might free the industry's entrepreneurial spirit. As a result of restructuring, the number of players interested in knowing the operational state of the grid has skyrocketed from a handful of vertically integrated utilities to hundreds of marketers, independent generators, regulators, and consultants. Currently, generation companies are making investment decisions about new plants, which cost hundreds of millions of dollars, based on very limited information about actual grid operation. This situation is almost guaranteed to produce some disastrous choices. New transmission lines may be needed, but how can governmental agencies and the public make informed decisions when information about actual grid operation is unavailable? New transmission technologies are being developed, but how can their manufacturers make informed business decisions about which technologies to pursue when they have limited means to determine need?

If data were available, third parties might quickly develop innovative informational products to meet the industry's needs. Third parties interested in selling to a market much larger than the traditional utility EMS market could develop many of the tool sets needed for analyzing large RTOs. Even with the limited data available today, third parties are offering some innovative grid analysis and visualization products. Increased availability of data might also allow for more effective independent oversight of grid operation. Currently, there is little oversight. Federal and state regulatory agencies do not have the tools or the data to effectively oversee grid activities, and because there is no access to these data, there is little incentive for third parties to develop the requisite tools.

Useful data might include transmission device status

information, real and reactive power flows for transmission facilities, voltages and frequencies at key points within the transmission network, along with more processed data such as ATC and LMP information. Given the current low cost of computer storage and the availability of high-speed data communication, dissemination of these data should be simple. For example, the posting of hourly snapshots of 5,000 flow values and 2,000 status values would require less than one megabyte of storage per day. Immediate public release of some data, such as generator offers, would not be appropriate.

As a result of increased concern about possible terrorist activity, public access to transmission system information has actually become substantially more restricted. For example, on October 11, 2001 FERC restricted public access to a substantial amount of energy facility data, including the FERC Form 715 data. This is unfortunate and, for the most part, unnecessary. Public access to a large amount of additional information is possible without jeopardizing either the physical security of the transmission system or legitimate proprietary concerns of grid participants. Without such data it will become increasingly difficult for market participants to effectively utilize the NTG. As a minimum, there is a need for an industry-wide discussion on what data can legitimately be made public and what data must remain proprietary, and on the best mechanism for the release of this data.

Regardless of whether data remain proprietary within RTOs or enter the public domain, key technologies for broad data coordination are digital communications, high-performance computing, computer mathematics, data management and mining, collaboration networks, information security, and operations analysis. Discussion of these subjects and partial templates for the needed R&D can be found in reports issued by the DOE and EPRI as part of the ongoing WAMS effort (DOE 1999). Much R&D for data coordination would draw upon and directly reinforce the evolving Reliability Information Network by which Regional Security Coordinators share grid information in near real time.

Technology Challenge #2:
Knowing the Limits of Safe Operation

Full use of transmission capacity means that the system will be loaded close to the "edge" of safe operation. In recent years, many cases have revealed this edge to be much closer than had been expected. Less dramatic yet of equal or greater importance are the many undocumented situations in which grid capacity has been significantly underutilized because lack of knowledge about real system limits resulted in overly conservative operation. Safe operating limits are defined by a multiplicity of system conditions that have become more difficult to anticipate, model, and measure directly.

Electrical conditions on the transmission system may not be fully known, and even if they were, their full implications might not be. It is not possible to anticipate and study all possible conditions, and the computer models used in studies are sometimes sufficiently unrealistic that they produce misleading results. Partial remedies would be to augment modeling results with measured data and to calibrate models against observed system behavior.

The challenge here is partly technical and partly institutional. On the technical side, the determination of safe operating ranges requires a variety of different inputs that are associated with a variety of different time frames, all of which are dependent on the accuracy of the underlying models and of the data provided to those models. The longest planning time frame is associated with operational limits set by planners weeks or months ahead. These usually include transient stability limitations, oscillatory stability limitations, and voltage stability limitations and are conditional on long-term forecasts of customer demand and overall power system resources. Because assumed conditions are seldom the same as actual operating conditions, the limits are intended to be sufficiently conservative that modest differences between predicted and actual operating conditions can be accommodated through later planning adjustments. These adjustments take place in a shorter time frame that supports planning for several hours to several days in advance. This shorter time frame permits more precise forecasts of pending system conditions, but it restricts the opportunity for in-depth analysis and the range of operational alternatives that can be considered. The planning and decision tools used in this time frame, though sometimes ad hoc, often provide market-critical information such as ATC to be communicated to market participants via the OASIS. Finally, in near-real time, system operators use the EMS to observe and assess the actual status of the power system. On-line tools, such as real-time power flow and contingency analysis, provide guidance for managing situations in which real-time conditions are substantially different from what was planned.

Ideally, the planning process insures that maximum transmission capability is available to the power market while system reliability is maintained. The challenge is that errors may arise at any point in the process. One problem, as noted above, is that the electrical conditions and their implications may not be fully known. The system must be observed in such a way that system operators receive timely and complete information. Complicating observation of the system is what is known as the issue of "seams" between areas of the grid. Currently in the U.S. there are approximately 140 different utility control areas and 20 higher-level security coordinators, each trying to monitor its portion of the grid. As EPRI (2001) notes, "each control entity is like its own sovereign nation as far as market and data practices go, and coordinating power transfers that extend beyond the borders of an entity entails complex technical tradeoff analyses consistent with how the grid actually responds to inter-regional power flows." Power flows easily between control areas, and, as illustrated by the example in Figure 3, the transactions and control actions in one or two areas can have grid-wide implications.

Another source of error in the planning process is that assumed conditions may differ widely from real-time conditions. If the planning limits are too high, or if some market-driven transfers become too heavy, the system may be in danger of widespread, cascading outages. In these circumstances, some market activities would have to be curtailed through actions such as TLR (transmission loading relief). Alternatively, planning limits may be too low or ATC results too conservative. In these cases, the transmission grid may be underutilized, with the market sending erroneous signals to adjust more generation or transmission than needed. One example is transmission line thermal limits, which in many markets are the limiting constraints on ATC. The amount of power that can be transferred along a line is highly dependent on ambient weather conditions. Yet fixed limits are used in most cases (sometimes these limits differ in winter and summer). Better estimation of these limits, perhaps coupled with real-time measurement of conductor temperature or sag, could result in a significant increase in ATC. The seams issue arises here as well because each security coordinator is simultaneously performing studies to determine transmission capability, usually without detailed knowledge of what its neighbors are doing.

A third source of error is flawed conceptual formulation

of the models that are used to predict power system behavior under highly stressed conditions. A common theme in the post-mortem analyses of major system disturbances is that the models did not correctly predict or replicate actual system behavior. One recent example is the near-voltage-collapse in the Pennsylvania-New Jersey-Maryland connection (PJM) during July 1999 (DOE 2000). Effective intervention by PJM operators averted a loss of load, in large part as a result of EMS technology that afforded unusually good real-time observation of grid voltages. Later analysis revealed substantial optimism in the assumed capabilities of many PJM generators to support system voltages (through reactive power generation) while producing specific levels of real power (megawatts). These findings parallel utility experience around the world: the actual capability and behavior of a thermal power plant may be radically different from that indicated by generator models or nameplates. This seems especially true of gas-fired turbines, which constitute almost all new plant construction. (It has been reported that some operators outside the U.S. take their plants to maximum output every hour, just to establish capability limits.) The emerging picture is that reserve generation capability for emergency use is much smaller than previously believed, and that financial considerations may encourage plant operations changes that compound the problem in ways that system planners are just now starting to recognize. This is one of several issues that the U.S. Department of Energy (DOE) has been monitoring through its Transmission Reliability Program.

A related issue associated with the NTG study is the need for better computer modeling of the interrelationships between electricity markets and the NTG. An accurate assessment of the cost impact of the NTG bottlenecks on market operation requires detailed, time-varying analysis (e.g., hour by hour) of an entire interconnected system. Since in some portions of the NTG the constraints are due to reactive/voltage problems, traditional, linear transportation-based models are not adequate. Such analysis could prove crucial to determining the optimal locations for expanded transmission capacity. Previously, such detailed analysis had been computationally prohibitive. However, faster computer processors and greater availability of parallel processing are rapidly removing these barriers. Development of the necessary computer models and algorithms for this analysis has also been hindered by lack of availability of the interconnect-wide data needed to perform such an analysis.

From a technical viewpoint, the immediate solution is to continue the incremental changes that have been taking place. These include developing enhanced real-time systems for measurement-based information, improved tools for system analysis and visualization, improved data communication between control centers and security coordinators, increased utilization of improved computer technology to move system limit calculations closer to real time, and increased feedback of system operational data to system planners to improve the calibration of models against observed system behavior. There is a significant need to improve our understanding of the fundamental behavior of the power system and the conditions or events that lead to system failure. Improved models are an essential element of this effort. Proactive federal involvement in the development of interconnect-wide models and tools could be quite helpful.

The solutions noted above neglect relevant institutional issues. Simply stated, in most markets there is a fundamental dichotomy between the commercial participants and the transmission managers who make the market possible. Unlike the commercial participants, the managers have no clear "pay for performance" mechanism for recovering their financial investments. The absence of such a mechanism has fostered a spiraling decline in staffing, priority, and overall resources given to system planning. Calibration of planning models and direct assessment of power system behavior should be integral to the planning process. The industry has a growing wealth of data to support this conclusion, not only from its EMS facilities but also from a host of sources including integrated phasor measurement systems and substation-based data recorders. Unfortunately, most of the utility staff with access to these data are too burdened by day-to-day tasks to use the data or the tools required to analyze the data. Repeated staff reductions have meant that this complex task has almost vanished from utility organizational charts. As highlighted in EPRI (2001), the linkages among markets, technology, and policy are fundamental and must be understood and adjusted to best effect.

Key technologies for this understanding are essentially the same as those noted for Technology Challenge #1. Special requirements include mathematical systems theory, signal analysis, operations analysis, and probabilistic methodology.

Technology Challenge #3:
Extending the Controllability of Network Flow

A higher degree of power flow control than is currently possible is a very attractive means to improve utilization of transmission resources. Conventional power-flow control devices include series capacitors to reduce line impedance, phase shifters, and shunt devices that are attached to the ends of a line to adjust voltages. A far higher degree of control is provided by HVDC transmission equipment and FACTS technology. The so-called NGH (a device, in which power electronics facilitate safe application of a conventional series capacitor) appears to be a precursor to FACTS technology.

Devices that improve flow control can be used individually or in combination to directly regulate power routing on the grid and to relieve dynamic problems that may limit grid utilization. Control of this sort is a very attractive alternative to the construction of new or stronger lines. This is not the whole story, however, because power system controls are subject to errors in the control law on which they are based or the models from which the control law is developed. (This is in contrast to the functional reliability of a new transmission line or power plant, which is almost synonymous with its hardware reliability.) Because of this vulnerability, the overall reliability of large-scale control systems cannot be assessed or assured by the straightforward and proven methods that are used in construction-based reinforcements to the grid. How, then, should the choice be made between controls and construction of new transmission capacity?

A full demonstration of controller reliability is rarely possible. It is always necessary to trade controller benefits against the risks associated with closing a high-power control loop around system dynamics that are not fully observed and not fully understood. Controller reliability must be assessed broadly, incorporating engineering judgment and sound practice. Uncertainty should be mitigated where possible, but this is often a slow and technically difficult process (Hauer & Hunt 1996). Whatever uncertainty cannot be mitigated should be accommodated in controller design and operation. All of these measures require that wide-area control systems be supported by wide-area information systems, and that the grid management infrastructure include an appropriate degree of technical expertise in control engineering (Hauer & Taylor 1998).

Wide-area control, whether using FACTS or less advanced technologies, offers many benefits to the next-generation national transmission grid. A recent FACTS installation in Brazil is especially noteworthy; it links two regional systems with an AC line plus two thyristor-controlled series capacitor (TCSC) units. Prior to this, a DC line would have been the inevitable and more expensive choice.

Here in the U.S., the installation by the New York Power Authority of a Convertible Static Compensator (CSC) FACTS device has increased the power transfers on the Utica-Albany power corridor by 60 MW in its initial phase, with a projected increase to 240 MW when Phase Two is completed in 2002. However, it is important to place these numbers in context. Overall, the peak electricity demand in New York State is approximately 30,000 MW, with approximately half the demand in upstate New York and the remainder in New York City and Long Island. The current import capability from the upstate region to the city and Long Island is approximately 4,500 MW, with another 2,000 MW coming from PJM. Therefore, the increase from the CSC device is approximately five percent of the current capacity, and about 1.5% of the peak New York City/Long Island load.

A proposal that complements the use of FACTS devices to achieve better network control is to break up the current Eastern and Western Interconnections into smaller, more manageable synchronous interconnections. These smaller interconnections (which could correspond to existing regional reliability councils) would be joined by HVDC ties; the size of the ties would match existing transmission transfer capabilities (De Steese and Dagle 1997). The use of HVDC between the interconnections would permit complete control of power flows between interconnections, completely eliminating long-distance loop flow. Loop flow would still be an issue within the interconnections, but their smaller size would make this flow easier to manage. Of course, such wide-scale dismantling of the Eastern and Western Interconnections would require major investment in new HVDC lines and could present a host of new, unforeseen technical problems.

A key challenge to the use of advanced technology to achieve better network control is that it is a "high tech" option entering the business environment for utilities, ISOs, and other grid managers, which today favors "low tech" investments. Advanced control technology is often characterized by high initial costs and ongoing maintenance and operation costs. In addition, use of HVDC or FACTS devices can result in higher power losses, with typical converter losses of one or two percent of flow. For most utilities, cost-benefit analysis currently favors doing nothing, letting new generation take care of the need, or investing in familiar passive AC devices. Outstanding issues to be addressed before advanced technologies can compete in such an environment include the need for operational experience, quantification of benefits, and resolution of impediments to reliable control in high-performance applications.

High-performance hardware for wide-area control is ready for use; conventional technologies have served local and regional needs for many years. Full use of wide-area control demands an improved infrastructure for wide-area information. WAMS, the information counterpart to FACTS control, is expressly designed to provide this infrastructure.

Technology Challenge #4:
Dealing with Operational Uncertainty

Providing reliable and economical electric power calls for two parallel efforts related to uncertainty. The first is to reduce uncertainty by means of information that is better and more timely than what is currently available. The second is to accommodate the residual uncertainty through the use of appropriate decision tools.

In 1996, two massive breakups of the western power system demonstrated the need for improved resources to deal with the unexpected. As noted above, data collected in real time at BPA's Dittmer control center contained subtle but definite indications of oscillatory instability for several minutes prior to the actual breakup on August 10. BPA operators also reported that hints of weak voltage support may have been present for much longer. Had there been means for converting these hints to unambiguous operator alerts, that breakup might have been avoided entirely.

Contradicting actual system behavior, later studies performed with standard WSCC models (adjusted to the conditions and events leading to the breakup) indicated that the system had excellent dynamic stability. Enhanced models, internally adjusted to match observed system behavior, are outwardly more realistic but still suspect. Modeling errors are one of many uncertainties that improved resources for grid management must accommodate.

Even if suitable planning models had been available, operating conditions preceding the August 10 breakup were far from nominal and had not been examined in system reliability studies. These studies are generally performed weeks to months in advance, and planners cannot anticipate all combinations of seemingly minor outages that may be part of the operation of a large power system. Planning uncertainty and its attendant risks can be mitigated in part if system capacity studies are performed with a much shorter forecasting horizon and based on reasonable extrapolations of current operating conditions. This approach calls for much broader real-time access to those conditions than any one regional control center now provides. The requisite computer tools are directly consistent with the framework envisioned for dynamic security assessment (DSA), however. This is also true of the measurement-based operator alerts mentioned earlier although the mathematics needed is quite different.

The combinatorial problem for longer-term planning remains especially formidable. The number of likely contingency patterns, already huge, is becoming even larger as the market seeks energy transactions across longer distances. Future practices may also represent model errors as contingencies. Even without this change, direct examination of each individual contingency pattern is not feasible. Contingency evaluation is a further challenge. Never a simple matter, it must now reflect new linkages between system reliability and market economics. Decisions must be rendered more rapidly than before despite increased uncertainty and sometimes increased risk.

Reducing and accommodating these uncertainties requires a broad, multi-faceted effort. Requisite technologies include:

- Improved real-time tools to examine power system signals for warnings of dangerous behavior. The more rapidly that operator intervention is initiated, the more likely that a blackout can be averted.

- Improved visualization, giving operators a bird's-eye view of the power system.

- Mathematical criteria, tools, and procedures for reducing and/or characterizing errors in power system models.

- Characterizations and probabilistic models for uncertainties in power-system resources and operating conditions.

- Probabilistic models, tools, and methodologies for collective examination of contingencies that are now considered individually.

- Cost models for quantifying the overall impact of contingencies and ranking them accordingly. It is essential that these models be realistic and suitable for use as standards for planning and operation of the overall transmission grid.

- Risk management tools, based on the above probabilistic models of contingencies and their costs, that "optimize" use of the electricity system while maintaining requisite levels of reliability.

Development of the technology noted above can likely be expedited through technology transfers from outside the power industry. Even so, there are special and difficult problems. The knowledge base for actual power system behavior, required both to define the subject technologies and obtain best value from their use, is not well evolved. The knowledge base and the technologies should develop together, in or close to a practical utility environment.

Furthermore, probabilistic planning is not just a smooth extrapolation of current practices. It requires new skills and practices. These practices must be developed, evaluated against those now in use, and then approved for use at the RTO level. These matters should be addressed at the earliest possible stage of technology development.

Technology Challenge #5: A Grid that Heals Itself

The interconnection of large power systems into still larger ones greatly increases the possibilities for widespread failure. Grid managers go to great lengths to anticipate and avoid such failure. However, at some level of complexity, anticipation and avoidance become too difficult or expensive.

A variety of lessons can be extracted from the 1996 breakups of the western power system. One of these lessons is that when prevention of system breakups becomes impractical, it is time to focus on minimizing the consequences. A triage approach has long been characteristic of grid operations; the operation of individual relays is a good example of removal of a small portion of the system to save the whole. On a broader basis, the use of under-frequency load shedding has been a very effective means of saving the grid from frequency decay at a cost of perhaps five or 10 percent of total load. Limited self-healing is also found in the use of automatic circuit-break reclosing after events such as the loss of a line from a lightning strike.

What is new since 1996 is a shift in emphasis from aggressive use of preventive control, accepting possible loss of some load, to consideration of "dynamic islanding" strategies that accommodate an occasional breakup while minimizing its impacts and assuring smooth restoration of electricity services. Dynamic islanding would involve:

- Emergency, possibly localized, controls that separate the power system into either predefined islands or dynamically defined islands as dictated by conditions that are sensed locally.

- Islanding options designed for minimal loss of service, given proper control assistance.

- Islanding and restoration as a continuous smooth process controlled by FACTS, HVDC, or automatic generation control.

- Conversion of some AC lines to DC. This change is particularly attractive for lines that would otherwise become stranded assets under the pressure of new generation installed close to major loads.

Avoiding just one catastrophic event would likely payback much of the investment cost of FACTS technology and the associated infrastructure. But there are both technical and institutional concerns. From a technical point of view, developing even limited islanding capability, let alone grid self-healing, is an immense challenge. During islanding, the two new islands will be simultaneously presented with a combination of potentially large initial generation/load imbalances and changes in line flows as existing tie flows are eliminated. Frequency regulation characteristics will also change due to the changes in total inertia. Although a dynamic islanding scheme has been implemented on the WSCC system, its use on the Eastern system would be much more involved because of the higher density of tie lines. From an institutional point of view, incentives would be needed to encourage the development of islanding schemes. It appears that only an ISO or RTO would in a strong business position to assume the costs given the geographical area involved.

Technology Challenge #6: More Power in Less Space

There are many reasons to seek power system equipment that requires a minimum of space. New rights of way for transmission lines are environmentally intrusive, difficult to route, and subject to a very slow approval process as local authorities are increasingly reluctant to approve projects that do not address local need. These problems tend to be less severe for underground transmission cables, but new routes or added space for underground cables may be impossible in highly urbanized environments like Chicago or New York City; costs inhibit the use of underground cables in less urban areas. Substations, generators, and transformers all benefit from having a smaller "footprint," especially if the equipment itself is smaller and more portable.

So how can we fit more power into a given space, or into even less space? Conventional solutions include "reconductoring" lines to carry more current at the same voltage, revising lines to operate at higher voltage (if possible), and converting AC lines to DC. The use of composite materials is a promising approach for reconductoring. Traditionally, overhead transmission lines have been constructed using aluminum conductors steel reinforced (ACSR) consisting of stranded aluminum about a stranded steel core. The aluminum carries the current, and the steel provides mechanical support. The limiting constraint for such lines is sag resulting from heating. A new approach for increasing conductor current capacity without increasing weight is to replace the steel core with a composite material, such as glass-fiber. Because the tensile strength of the glass is up to 250 percent of the strength of steel, the composite conductors are lighter and stronger and could have higher current capacity. Reduction in sag allows tighter spacing of conductors, which reduces magnetic fields and might mean that new conductors could be added in existing rights of way.

A complementary approach to increasing the available capacity of existing AC lines is to dynamically determine the actual conductor limits. The thermal capacity of an overhead line is highly dependent on ambient conditions; there is more power-transfer capacity when the line is being operated in cold, windy conditions than when it is operating in hot, calm weather. Approaches to dynamically determining conductor limits include either direct measure of conductor temperatures or use of a differential global positioning system (GPS) to directly measure the sag of critical spans.

For higher-voltage lines, limits are usually based on "loadability" constraints rather than thermal limits. The loadability of a line that cannot be operated close to its thermal limit can often be improved by compensating devices or full FACTS control. Another promising though relatively conventional technology is compact transmission lines that are reconfigured to carry more power (at the expense of increased losses).

An alternative to overhead lines is buried cables. Several different cable designs can be used; oil-impregnated paper-insulated pipes are the most common. A key advantage of underground cables is that they usually face little public opposition. Also, the closer spacing of the conductors results in greatly reduced electromagnetic fields (EMFs) because of phase cancellation. Finally, underground cables are not subject to weather and thus may be more reliable than aboveground lines. The key disadvantage of buried cables is cost. With cost ratios of up to ten times for rural high-voltage lines, it is nearly always more economical to build overhead lines unless one is in an urban area. Also, the length of AC cables is limited by their relatively high capacitance; uncompensated cables may be limited to perhaps 25 miles. Finally, over the long term, underground cables may not be as reliable as overhead lines because it takes substantially longer to locate and correct problems with buried lines.

Truly strategic improvements in compactness call for new technologies like supercapacitors, transformerless HVDC, and cryogenically enhanced devices. Cryogenic operation (i.e., operation at unusually low temperatures, which may or may not be low enough to achieve superconductivity) reduces or eliminates resistance in an electrical device and thereby allows a several-fold in increase in its power-handling capacity. This benefit can be exploited either as increased capacity within given size and weight constraints or as equivalent performance in a much smaller and lighter package. However, cryogenic devices also have disadvantages. For example, some super-conducting devices operate with extremely high currents and thus radiate very intense magnetic fields. As a general rule, the introduction of cryogenic cooling adds complexity to a device, so a utility using cryogenic devices would have to hire employees with the specific skills to maintain these devices. Cryogenic devices also generally require long cool-down times, up to a week or more for some such as super-conducting magnetic energy storage (SMES) and large transformers. Certain maintenance and repair procedures may require warming the devices up to ambient temperature, which takes a similar amount of time. This characteristic may be an unacceptable operational constraint.

Cryogenic devices now include cables, transformers, current limiters, switches, generators, and energy storage devices (SMES). These devices are at stages of development ranging from working prototypes to a few commercially successful products. The underlying base technologies are the subject of active research, and the technical feasibility of cryogenics in general is increasing steadily. As with FACTS, the chief impediment to practical deployment is the initial investment.

Another partial solution to current difficulties with obtaining new rights of way is to utilize non-traditional transmission paths, such as submarine cables. One such project currently under consideration, known as the Neptune Project, seeks initially to connect 345-kV substations in Brooklyn and Long Island NY with a 345- kV substation in northern New Jersey via two 600-MW HVDC cables buried in trenches on the Atlantic Ocean floor. Subsequent phases seek to link New York City with a substation in New

Brunswick, Canada using a 1,200-MW submarine HVDC; cables added later would join Boston and other New England locations. Another project under consideration seeks to link Ontario, Canada to either Ohio or Pennsylvania using several HVDC cables under Lake Erie. Given the large number of urban load centers located on the oceans or Great Lakes, the commercial success of one such submarine HVDC project could lead to many more. The advantage of such an installation is that no eminent domain authority is needed to obtain the water rights of way, and the relatively small land-based converter stations that are required can be located to bring power directly into urban load centers. Due consideration must be given to avoid harming the aquatic environment, with the cables routed to avoid active fishing and sensitive environmental areas.

Technology Challenge #7:
Assessing New Technologies Using Life-Cycle Analysis

Investments in new technologies can be both necessary and dangerous. Investments in the wrong technologies can lead to disaster. Timely new ways to reduce costs and improve performance are essential to business survival. Utilities tend to be very cautious in investing in new technology.

One reason for their caution is that the actual merits of any new technology can be difficult to estimate in advance. New or advanced technologies are very likely to have hidden costs (and may also have hidden benefits). Some technologies are "fragile," requiring significant engineering design or unforeseen maintenance. Others are "intrusive" in that their use calls for major changes in associated technologies and methods. Still others produce long-term environmental problems, such as the disposal of hazardous materials used in their construction. And some technologies might fail in a catastrophic manner that endangers human health and safety.

All utilities are well aware of these possibilities, but few individual utilities have the resources to assess them. Suitable resources can be assembled on a collaborative basis, but a suitable assessment methodology must also be developed.

Full assessment of new technologies calls for a life-cycle analysis that considers all costs and all benefits, with suitable consideration of regulatory constraints and other external or uncertain factors. To be inclusive, life-cycle analysis should start with production of the technology and consider impacts upon the economy, health and safety, the natural environment, and other elements of the public good. The analysis continues from this point through all expected uses of the product to its eventual recycling or disposal by other means.

The electric power industry seldom makes equipment acquisitions using such thorough analysis. The industry will also argue, reasonably, that it cannot afford in-depth consideration of all aspects of the public good in everyday business decisions. However, the norm for much equipment procurement is to accept the minimum bid. This practice has already populated the national grid with a large amount of energy-inefficient equipment. The practice of life cycle cost optimization should at least consider the full range of tradeoffs, comparing benefits with the total cost of ownership for the life of the equipment. Life-cycle costs include acquisition costs as well as costs of capital, energy, operations, maintenance, and disposal.

Technology Challenge #8: The Intelligent Energy System

Information is the crosscutting issue in all transmission grid technology challenges. WAMS and FACTS share an underlying vision of an Intelligent Energy System (IES) in which "intelligent" planning, design, control, and operation of system assets are the primary means for meeting energy demands. An IES might well involve coordinated operation of the electrical and gas energy systems, with the gas system providing virtual storage for electrical energy. The IES would certainly draw upon FACTS technology for the routing of electrical power and upon dispersed assets such as distributed generation, energy conservation, direct or indirect load control, and renewable energy sources. WAMS is a critical element in the information infrastructure needed to make the IES possible and to insure power system reliability.

The vision of an IES extends beyond FACTS and perhaps beyond WAMS. Additional elements include protective relay systems that "adapt" to widely variable power flows, diagnostic tools to reduce human error during system maintenance, enhanced information tools for emergency management, and "intelligent" data miners that sift operating records for evidence of needed maintenance. Some specific examples, extracted from much more detailed treatments in PNNL (1999), are presented below.

Protective Controls—Relay Coordination

Containing a sizable disturbance usually requires appropriate action by several relays. Communication among the relays is often indirect, through the power system itself. Effectively designed direct communication among relays would make coordination more reliable from the hardware perspective. Relays, like transducers and feedback controllers, are signal-processing devices that have their own dynamics and modes of failure. Some relays sense conditions (like phase imbalance or boiler pressure) that power-system planners cannot readily model. At present, there are few engineering tools for coordinating wide-area relay systems.

Large power systems are sometimes operated in ways that were not foreseen when relay settings were established. It is not at all apparent that fixed relay settings can accommodate the increasingly busy market or, even more difficult, the islanding that has been seen recently in North America. It may be that relay-based controls, like feedback controls, will need some form of parameter scheduling to cope with such variability. The required communications could be highly vulnerable from a security standpoint, however, so precautions against the growing threat of "cyber attack" would be needed.

Several recent grid events suggest that there are still questions to be resolved regarding the basic strategy or economics of bus protective systems (PNNL 1999). In the western system breakup of December 14, 1994, it appears that "bus geometry" forced an otherwise unnecessary line trip at the Borah substation in Idaho and led directly to the system breakup. Bus geometry was also a factor when all transmission to San Francisco was lost on December 8, 1998. Following routine maintenance at the San Mateo substation, a breaker was closed while protective grounds were still attached. The resulting fault tripped all lines to the San Mateo bus because a differential relay system had not been fully restored to service. An appropriate diagnostic tool would have indicated this condition and warned that the grounds had not been removed.

Emergency Management—the Northeast Ice Storm of 1998

Emergency management resources of the Northeast Power Coordinating Council were severely tested when a series of exceptionally severe ice storms struck large areas in New York, New England, Ontario, Quebec, and the Maritime provinces between January 5 and 10, 1998. The worst freezing rains ever recorded in that region deposited ice up to three inches thick. Resulting damage to transmission and distribution was severe (more than 770 towers collapsed).

The event resulted in some valuable lessons regarding system restoration. Emergency preparedness, cooperative arrangements among utilities and with civil authorities, integrated access to detailed outage information, and an innovative approach to field repairs were all found to be particularly valuable. The disturbance report mentions that information from remotely accessible, microprocessor-based fault locator relays was instrumental in quickly identifying and locating problems. Implied in the report is that the restoration strategy amounted to what mathematicians call a "stochastic game," in which some risks were taken in order to make maximum service improvements in the least time— and with imperfect information about system capability.

Technology Challenge #9:
Physical and Cyber Security of the Transmission Grid

Given the recent increased awareness of the possibility of terrorist activity, it seems especially pressing to address the physical security of the NTG. (This paper focuses on the transmission system only and does not address the physical security of individual generation stations.) We consider transmission security in relation to the risk of physical destruction of system elements and concerns about cyber security.

In relation to concerns about physical destruction, the blessing and the curse of the transmission grid is its immense size. In the U.S. there are currently more than 150,000 miles of transmission lines that are 230 kV or higher, and there are many tens of thousands more miles at lower voltage levels. In both the Eastern and Western Interconnects, there are tens of thousands of individual transmission lines and many thousands of individual high-voltage transformers. The curse is that such a system is impossible to "secure"; there is no effective means to prevent a determined group of individuals from destroying a portion of the grid. But the blessing is that they could destroy only a minuscule portion. In addition, any destruction aimed of individual towers would have temporary effects. Given the regular occurrences of tornadoes, hurricanes, ice storms, and earthquakes, the transmission system has been designed to take its share of individual hits and continue to function. And the utility industry is quite adept at quickly repairing the damage done from such natural occurrences. It would be very difficult for even a large, well-organized group to duplicate the physical damage done by even a moderate ice storm.

The issue is whether a major disruption could be caused if various key grid facilities, such as electric substations or rights of way with many individual circuits, were selectively targeted. The answer is "yes" if enough key facilities were destroyed. But the impacts would likely be temporary because transmission lines could relatively quickly be rerouted around most substations. Some equipment (e.g., transformers) would be vulnerable and difficult to replace. The destruction of multiple transmission stations by a knowledgeable saboteur with a highly organized attack could result in substantial damage and long-term blackouts.

Another concern is security of information systems or cyber security. The increasing reliance of the electric power industry on communications and control systems together with the remarkable advance of electronic intrusion technologies and techniques make the restructuring utility industry particularly vulnerable to disruptions resulting from inadequate safeguards and security capabilities. More points of entry into command and control systems will become available to potentially hostile individuals or organizations. Many of these entry points will differ from the points of access previously established to serve a vertically integrated utility industry. The advent of real-time power dispatching coupled with competition in retail power markets and many other challenges of operating in a restructured industry environment will greatly reduce the safety margins currently maintained by electric utilities. The utility system of the future could become much more vulnerable to corruption by skilled electronic intrusion from both inside and outside. A primary, emerging need in the utility industry is for development of new guidelines, policies, and standards for the selection and implementation of cost-effective security measures (EPRI 1996).

The protection of critical civilian infrastructure has been a national focus since the mid-1990s with the formation of the Presidential Commission on Critical Infrastructure Protection (PCCIP) in July 1996 and the Presidential Decision Directive 63 (PDD-63) on Critical Infrastructure Protection issued in May 1998. DOE is responsible for the electric power sector and the natural gas and oil production and storage sectors and has formally designated NERC and the National Petroleum Council (NPC) as liaison organizations. In April 2001, NERC published a white paper: Approach to Action for the Electricity Sector that outlines the electric power industry's plans for security against physical and cyber attack.

For lack of a clear "business case," new technology investments often involve more financial risk than any single utility (or new ISO) can accept. Motivating these investments will require some combination of definitive national policy along with market models for investment planning.

MEETING THE TECHNOLOGY CHALLENGES

Institutional Issues

A formidable number of institutional issues hinder timely identification, development, and introduction of new technologies. Today's utilities are understandably reluctant to fund R&D that is not promptly and directly beneficial to them. Likewise, utility suppliers will not fund new transmission technology research if there is not a reasonable likelihood of an adequate return on the investment. Contrary to the premises under which EPRI was established separately from the DOE national laboratories, it is now difficult for EPRI to act as the coordinating umbrella organization for long-term R&D in the public interest. Much of the work produced by EPRI is essentially unavailable to the many nonmembers. The following "out-of-the-box" solutions should be considered:

- Apply a user fee to all institutions that engage in energy business. This fund would be used exclusively for energy R&D in the public interest, and all R&D results would be fully available to all energy business institutions.

- With suitable oversight provisions, disperse the above R&D fund through a DOE entity or a new public-service arm of EPRI (all institutions that engage in energy business would be members). It might be preferable to coordinate and consolidate these activities through a new umbrella organization for energy R&D.

- Engage industry experts in mentoring R&D and in-the-field assessments that are needed to close the gap between the development of new technology and its actual deployment for operational use.

Effective Utilization of Federal Resources

The federal government is very involved in the national grid through the federal utilities, including the TVA, the Power Marketing Administrations (PMAs), various elements of the U.S. Army Corps of Engineers and of the U.S. Bureau of Reclamation, as well as other entities. Collectively, the federal utilities operate "backbone" facilities for a large portion of the North American power system.

The federal government is also the ultimate steward for the staff skills, knowledge, and operational infrastructure of the federal utilities. These utilities are unique national resources of great value. They are immediately available to reinforce energy reliability in the public interest, a role in which they have long been a mainstay. Consideration should be given to the following ways to better utilize this resource:

- Fully engage the federal utilities as advisors and/or researchers in ongoing federal efforts to meet national energy needs.

- Draw on the federal utilities for field testing and operational assessment of new or prototype technologies. Give special attention to critical enabling technologies that have not drawn sufficient commercial interest to assure their timely evaluation and refinement.

- Identify critical resources provided by the federal utilities and integrate these resources into the national laboratory system. Support could be provided through a consortium arrangement among the national laboratories and federal utilities.

- Take immediate steps to establish a productive dialogue among all members of the proposed consortium and safely archive their collective institutional knowledge for future use.

Effective Utilization of Academic Resources

The electricity industry may be underutilizing the R&D potential of American universities. However, a contrary view holds that industry needs are primarily in development and that universities lack both the mission and staff continuity to proceed past the initial research phase. The proper relationship between university and industry has not been determined and should not be regarded as fixed. What is clear is that the dialogue between universities and the electricity industry is weaker than is the university-industry relationship in most other industries and that few universities have the direct industry involvement or the "institutional culture" that is needed for practical technology development in this area. Changing this situation might lead to a university system closer to the European model, in which many academics are part-time industry employees. Fundamental changes in the relationship between university and industry have many ramifications, and an open discussion of the matter would be timely.

One bright spot is the growing trend toward cooperative university/industry research centers. These centers seek to bridge university-industry gaps by directly involving industry in university research projects. This partnership helps projects maintain a degree of focus on problems currently facing the industry. The challenge for the universities involved in such centers is to demonstrate to their industrial members that membership fees represent money well spent.

In addition, there are growing numbers of faculty members involved in start-up companies in the power area. Universities nationwide are seeing a need to foster economic development in their local regions and states and to facilitate the transfer of university expertise and research to industry. Although a number of mechanisms exist to meet these goals, encouraging faculty with innovative ideas to form start-up companies is particularly promising as much of the country's innovation arises from entrepreneurial activity by small companies.

ADVANCED TECHNOLOGY
RELATED RECOMMENDATIONS

Sustainable solutions require balances between generation, transmission, and demand; planning and operations; profit and risk; the roles of public and private institutions; and market forces and the public interest. The strategic need is not just for advanced technology in the laboratory but also for an infusion of improved technology at work in the power system. The chief impediments to this are institutional; they can be resolved through a proactive national consensus regarding institutional roles. Until this consensus is achieved, the lack of linkage between technology and policy may be a disruptive force in continued development of the national transmission grid and the broad infrastructures that it serves.

Listed below are several specific recommendations to address national transmission grid issues by creating a framework to accelerate the development and deployment of appropriate advanced technologies.

- Establish incentives for both private and public sector investment in RD&D. While federally funded basic research is important, ultimately it is the commercial sector that should move technology from the research lab to the marketplace. There is a need to accelerate the transition process within the electricity industry to reduce uncertainties regarding the future structure of generation and transmission markets since such uncertainties greatly impede investment in RD&D.

- Develop performance metrics for the national transmission grid where performance measures can be used to determine minimum planning and operational standards. These outcome- based measures would be consistent with the goals of the national transmission grid where issues such as serving the public good, promoting the economy, and ensuring national security can be balanced against the profit motivation associated with individual companies engaged in the electricity sector. Such a framework provides an incentive for private and public funding to research, develop, and deploy advanced technology because the linkage between enhanced performance associated with advanced technologies can be mapped to specific goals, with emphasis upon those technologies that cost effectively address poor performance according to these established measures.

- Apply a user fee to all institutions that engage in energy business. This fund will be used for energy R&D that is performed in the public interest, and all R&D results will be available to all institutions that engage in energy business.

- Stimulate the research, development, testing, and deployment of cost-effective technologies that allow greater capacity in existing right-of-ways. This includes passive reinforcement (e.g., advanced conductors and transmission configurations), active reinforcements (e.g., FACTS, HVDC, energy storage, and non-transmission technologies), and advanced information resources and controls that facilitate best use of transmission resources while ensuring system reliability.

- Promote programs that provide opportunities for energy consumers to manage their distributed energy resources (generation, storage, and load) in response to competitive market forces, including increased price visibility, demand-side participation in energy and ancillary services markets, and removal of technical and institutional barriers to distributed energy resources.

- Provide a forum for an industry-wide discussion to reach consensus on information access and dissemination issues. Certain information on system operations should be made broadly available to encourage markets to function, yet other information may be proprietary or sensitive.

- Establish security standards (both physical and cyber) for protecting the national transmission grid from attacks of malicious intent. Such standards should be derived from ongoing research in recognition of the evolving threat against the National critical civilian infrastructures.

- Draw upon the Federal utilities as a uniquely available and competent technical platform for inclusion in an expanded national RD&D infrastructure. Identify resources (facilities, staff, software, etc.) that do or could provide essential support to planning, development, and operation of the North America power system.

- Engage industry experts in the mentoring of R&D efforts and in the field assessments (demonstration projects) that are needed to close the gap between the development of new technology and its actual deployment for operational use. Give special attention to "critical path" enabling technologies that have not drawn sufficient commercial interest to assure their timely evaluation and refinement.

References

De Steese, J.G. and J.E. Dagle. 1997. "Electric utility system applications of fast-acting energy storage as illustrated by SMES." *Int. J. of Global Energy Issues*, Vol. 9, No. 3: 113-127.

EPRI. 1996. *Assessment of Information Assurance for the Utility Industry.* Report PNWD-2393, prepared by Battelle Northwest for the Electric Power Research Institute under Research Project 8024-01, Dec. 1996.

EPRI. 1997. *Powering Progress — The Electricity Technology Roadmap Initiative.* EPRI Technical Report, May.

EPRI. 2001. *The Western States Power Crisis: Imperatives and Opportunities.* June 25. http://www.epri.com/journal/details.asp?doctype=news &id=161

Fleishman, B.J., 1997. *Emerging Liability Issues for the New Electric Power Industry.* 1997 IBC Conference on Ensuring Electric Power Reliability in the Competitive Marketplace, San Francisco, CA, September 29-30.

Hauer, J.F. and J.R. Hunt, in association with the WSCC System Oscillations Work Groups. 1996. *Extending the Realism of Planning Models for the Western North America Power System.* V Symposium of Specialists in Electric Operational and Expansion Planning (SEPOPE), Recife (PE) Brazil, May 19-24.

Hauer, J.F. and C.W. Taylor. 1998. "Information, Reliability, and Control in the New Power System." *Proceedings of the 1998 American Control Conference*, Philadelphia, PA., June 24-26.

Lionberger, J. and L. Duke. 2001. "3D Modeling Boosts Transmission Capacity," *Electric Light & Power*, July.

North American Electric Reliability Council. 2001. *Working Group Forum on Critical Infrastructure Protection — An Approach to Action for the Electricity Sector*, April 5. Available at http://www.nerc.com/ ~filez/cipfiles.html

Oak Ridge National Laboratory. 1991. *Maintaining Electric Power System Performance: Preparing for the Year 2020, Research Needs.* Oak Ridge National Laboratory Report ORNL-6678, August.

Oak Ridge National Laboratory. 1991. *DOE Workshop on Real-Time Control and Operation of Electric Power Systems.* Proceedings of a Department of Energy Conference at Denver, CO. CONF-9111173. Nov. 19-21.

PA Consulting Group. 2001. *The Future of Electric Transmission in the United States*, January.

Pacific Northwest National Laboratory. 1999. *Review of Recent Reliability Issues and System Events.* PNNL technical report PNNL-13150, prepared for the U.S. Department of Energy Transmission Reliability Program by the Consortium for Electric Reliability Solutions (CERTS), December.

Roman, A.J. 1999. *Legal Responsibility for Reliability in the New Competitive Electricity Markets in Canada: Who Do I Sue if the Lights Go Out.* IEEE/PES Summer Meeting Plenary Session, "Reliability in the New Market Structure." Edmonton, Canada, July 18-22.

Scherer, H., 1999. "Reliability vs. Technology: A Race Against Time," IEEE Computer Applications in Power, vol. 11 no. 4, pp. 10-12, October.

U.S. Department of Energy. 2000. *Report of the U.S. Department of Energy's Power Outage Study Team— Findings from the Summer of 1999.* March. Available at website http://tis.eh.doe.gov/post/

U.S. Department of Energy [else Hauer, J.F., W.A. Mittelstadt, W.H. Litzenberger, C. Clemans, D. Hamai, and P. Overholt]. 1999. *Wide Area Measurements for Real-Time Control and Operation of Large Electric Power Systems—Evaluation and Demonstration of Technology for the New Power System,* Prepared for U.S. Department of Energy Under BPA Contracts X5432-1, X9876-2; January 1999. This report and associated attachments are available on compact disk.

U.S. Department of Energy. 1998. *Maintaining Reliability in a Competitive US Electricity Industry.* Report of the United States Department of Energy Electric System Reliability Task Force, September 29, 1998. Available on the Internet at http://vm1.hqadmin.doe.gov:80/seab/

U.S. Department of Energy, 1998. *Technical Issues in Transmission System Reliability.* Report of the United States Department of Energy Electric System Reliability Task Force, May 12, 1998. Available on the Internet at http://vm1.hqadmin.doe.gov:80/seab/

U.S. Department of Energy. 1980. *The National Power Grid Study.* U.S. Department of Energy Report DOE/ER-0056/3. January.

U.S. Department of Energy, undated. *National Energy Strategy: Powerful Ideas for America.* DOE/A-082P, Washington, D.C.

U.S. Federal Energy Regulatory Commission. 2001. Docket PL02-1-000, October 11.

U.S. Office of Technology Assessment. 1990. *Physical Vulnerability of Electric Systems to Natural Disasters and Sabotage.* Office of Technology Assessment Report. Government Printing Office 052-003-01197-2.

APPENDIX A: LIST OF NEW TECHNOLOGY EQUIPMENT TO REINFORCE THE TRANSMISSION GRID

The transmission system of the future must not only have increased capacity to support the market demand for energy transactions, it must also be flexible to adapt to alterations in energy-delivery patterns. These patterns change at various time scales: hourly, daily, weekly, and seasonally. The transmission system must also adapt to delivery patterns dictated by the evolving geographical distribution of load and generation. As generation planning and dispatch decision making are placed in the hands of organizations other than utilities, new technologies that afford transmission planners a wider range of alternatives for deployment of power become more attractive.

This appendix lists some of the newer hardware technologies that are being researched and deployed to reinforce grid operations. The range of potential technologies is enormous. This appendix is limited to the hardware technologies that are most directly applicable to grid operations; the list presented is not exhaustive. Software technologies are discussed in the body of the paper and are not addressed here.

The appendix organizes hardware technologies into the following categories:

- Passive reinforcing equipment,

- Active reinforcing equipment, and

- Real-time monitoring equipment.

Within each category, we list the relevant technologies and summarize the primary objective, benefits, barriers to deployment, and commercial status of each.

Passive Reinforcing Equipment

This section discusses the potential impacts of new technologies associated with AC transmission lines and related equipment (transformers, capacitors, switch gear, etc.). This category includes the increased value (capacity per unit cost of installation, operation, and maintenance) obtained through new conductor materials and transmission line configurations as well as the flexibility gained to reconfigure the transmission system through greater modularity of transmission equipment.

Passive AC devices constitute by far the majority of the existing network. Though new lines will certainly be needed to reinforce the grid, the siting of these lines will continue to be a major challenge. Getting the most out of existing rights of way minimizes the need for new lines and rights of way and can minimize the societal concerns associated with visual pollution and high-energy EMFs.

Conductors

Advances in conductor technology fall into the areas of composite materials, and high-temperature superconductors.

High-Temperature Super-Conducting (HTSC) Technology:
The conductors in HTSC devices operate at extremely low resistances. They require refrigeration (generally liquid nitrogen) to super-cool ceramic superconducting material.

Objective: Transmit more power in existing or smaller rights of way. Used for transmission lines, transformers, reactors, capacitors, and current limiters.

Benefits: Cable occupies less space (AC transmission lines bundle three phase together; transformers and other equipment occupy smaller footprint for same level of capacity). Cables can be buried to reduce exposure to EMFs and counteract visual pollution issues. Transformers can reduce or eliminate cooling oils that, if spilled, can damage the environment. The HTSC itself can have a long lifetime, sharing the properties noted for surface cables below.

Barriers: Maintenance costs are high (refrigeration equipment is required and this demands trained technicians with new skills; the complexity of system can result in a larger number of failure scenarios than for current equipment; power surges can quench (terminate superconducting properties) equipment requiring more advanced protection schemes).

Commercial Status: A demonstration project is under way at Detroit Edison's Frisbie substation. Four-hundred- foot cables are being installed in the substation. Self-contained devices, such as current limiters, may be added to address areas where space is at a premium and to simplify cooling.

Below-Surface Cables: The state of the art in underground cables includes fluid-filled polypropylene paper laminate (PPL) and extruded dielectric polyethylene (XLPE) cables. Other approaches, such as gas-insulated transmission lines (GIL), are being researched and hold promise for future applications.

Objective: Transmit power in areas where overhead transmission is impractical or unpopular.

Benefits: The benefits compared with overhead transmission lines include protection of cable from weather, generally longer lifetimes, and reduced maintenance. These cables address environmental issues associated with EMFs and visual pollution associated with transmission lines.

Barriers: Drawbacks include costs that are five to 10 times those of overhead transmission and challenges in repairing and replacing these cables when problems arise. Nonetheless, these cables represent have made great technical advances; the typical cost ratio a decade ago was 20 to one.

Commercial Status: PPL cable technology is more mature than XLPE. EHV (extra high voltage) VAC and HVDC applications exist throughout the world. XLPE is gaining quickly and has advantages: low dielectric losses, simple maintenance, no insulating fluid to affect the environment in the event of system failure, and ever-smaller insulation thicknesses. GILs feature a relatively large-diameter tubular conductor sized for the gas insulation surrounded by a solid metal sleeve. This configuration translates to lower resistive and capacitive losses, no external EMFs, good cooling properties, and reduced total life-cycle costs compared with other types of cables. This type of transmission line is installed in segments joined with orbital welders and run through tunnels. This line is less flexible than the PPL or XLPE cables and is, thus far, experimental and significantly more expensive than those two alternatives.

Underwater application of electric cable technology has a long history. Installations are numerous between mainland Europe, Scandinavia, and Great Britain. This technology is also well suited to the electricity systems linking islands and peninsulas, such as in Southeast Asia. The Neptune Project consists of a network of underwater cables proposed to link Maine and Canada Maritime generation with the rest of New England, New York, and the mid-Atlantic areas.

Advanced Composite Conductors: Usually, transmission lines contain steel-core cables that support strands of aluminum wires, which are the primary conductors of electricity. New cores developed from composite materials are proposed to replace the steel core.

Objective: Allow more power through new or existing transmission rights of way.

Benefits: A new core consisting of composite fiber materials shows promise as stronger than steel-core aluminum conductors while 50 percent lighter in weight with up to 2.5 times less sag. The reduced weight and higher strength equate to greater current carrying capability as more current-carrying aluminum can be added to the line. This fact along with manufacturing advances, such as trapezoidal shaping of the aluminum strands, can reduce resistance by 10 percent, enable more compact designs with up to 50 percent reduction in magnetic fields, and reduce ice buildup compared to standard wire conductors. This technology can be integrated in the field by most existing reconductoring equipment.

Barriers: More experience is needed with the new composite cores to reduce total life-cycle costs.

Commercial Status: Research projects and test systems are in progress.

Transmission Line Configurations

Advances are being made in the configuration of transmission lines. New design processes coupled with powerful computer programs can optimize the height, strength, and positioning of transmission towers, insulators, and associated equipment in order to meet engineering standards appropriate for the conductor (e.g., distance from ground and tension for a given set of weather parameters).

Tower Design Tools: A set of tools is being perfected to analyze upgrades to existing transmission facilities or the installation of new facilities to increase their power-transfer capacity and reduce maintenance.

Objective: Ease of use and greater application of visualization techniques make the process more efficient and accurate when compared to traditional tools. Traditionally, lines have been rated conservatively. Careful analysis can discover the unused potential of existing facilities. Visualization tools can show the public the anticipated visual impact of a project prior to commencement.

Benefits: Avoids new right-of-way issues. The cost of upgrading the thermal rating has been estimated at approximately $7,000 per circuit mile, but reconductoring a 230-kV circuit costs on the order of $120,000 per mile compared with $230,000 per mile for a new steel-pole circuit (Lionberger and Duke 2001).

Barriers: This technology is making good inroads.

Commercial Status: Several companies offer commercial products and services.

Six-Phase and 12-Phase Transmission Line Configurations: The use of more than three phases for electric power transmission has been studied for many years. Using six or even 12 phases allows for greater power transfer capability within a particular right of way, and reduced EMFs because of greater phase cancellation. The key technical challenge is the cost and complexity of integrating such high-phase-order lines into the existing three-phase grid.

Modular Equipment

One way to gain flexibility for changing market and operational situations is to develop standards for the manufacture and integration of modular equipment.

Objective: Develop substation designs and specifications for equipment manufacturers to meet that facilitate the movement and reconfiguration of equipment in a substation to meet changing needs.

Benefits: Reduces overall the time and expense for transmission systems to adapt to the changing economic and reliability landscape.

Barriers: Requires transmission planners and substation designers to consider a broad range of operating scenarios. Also, developing industry standards can take a significant period, and manufacturers would need to offer conforming products.

Commercial Status: Utilities have looked for a certain amount of standardization and flexibility in this area for some time; however, further work remains to be done. National Grid (UK) has configured a number of voltage-support devices that use modular construction methods. As the system evolves, the equipment can be moved to locations where support is needed (PA Consulting Group 2001).

Universal Transformer: A single, standardized design capable of handling multiple voltage transformations in the mid ranges of 161/230/345/500 kV on a switch-selectable basis. Added features might be high portability, to facilitate emergency deployment from a "strategic reserve" of such transformers, plus the accommodation of high phase order transmission lines.

Exotic Transmission Alternatives

The following technical approaches have been proposed to reduce losses, increase capacity, and/or address situations where traditional energy transport mechanisms have shortcomings. In all cases, test configurations have been developed, but commercial implementations have yet to emerge.

Power Beaming (Wireless Power Transmission): Power beaming involves the wireless transmission of electric energy by means of either laser or microwave radiation. Near-term applications include transmission of electric energy for space applications (e.g., to orbiting satellites) from either a terrestrial- or space-based power generation platform. Other applications that have been studied include supporting human space exploration (e.g., lunar or Mars missions). Future applications might involve the beaming of energy from orbiting or even lunar-based solar power generators to terrestrial receivers, but to date the economics of such a system have remained elusive; proponents of such systems believe that they can be competitive within 15 to 25 years.

Ultra-High Voltage Levels: Because power is equal to the product of voltage times current, a highly effective approach to increasing the amount of power transmitted on a transmission line is to increase its operating voltage. Since 1969, the highest transmission voltage levels in North America have been 765 kV, (voltage levels up to 1,000 kV are in service elsewhere). Difficulties with utilizing higher voltages include the need for larger towers and larger rights of way to get the necessary phase separation, the ionization of air near the surface of the conductors because of high electric fields, the high reactive power generation of the lines, and public concerns about EMFs.

Active Reinforcing Equipment
Transmission System Devices

Implemented throughout the system, these devices include capacitors, phase shifters, static-var compensators (SVCs), thyristor-controlled series capacitors (TCSC), thyristor-controlled dynamic brakes, and other similar devices. Used to adjust system impedance, these devices can increase the transmission system's transfer capacity, support bus voltages by providing reactive power, or enhance dynamic or transient stability.

HVDC: With active control of real and reactive power transfer, HVDC can be modulated to damp oscillations or provide power-flow dispatch independent of voltage magnitudes or angles (unlike conventional AC transmission).

Objective: HVDC is used for long-distance power transport, linking asynchronous control areas, and real-time control of power flow.

Benefits: Stable transport of power over long distances where AC transmission lines need series compensation that can lead to stability problems. HVDC can run independent of system frequency and can control the amount of power sent through the line. This latter benefit is the same as for FACTS devices discussed below.

Barriers: Drawbacks include the high cost of converter equipment and the need for specially trained technicians to maintain the devices.

Commercial Status: Many long-distance HVDC links are in place around the world. Back-to-back converters link Texas, WSCC, and the Eastern Interconnection in the US. More installations are being planned.

FACTS Compensators: Flexible AC Transmission System (FACTS) devices use power electronics to adjust the apparent impedance of the system. Capacitor banks are applied at loads and substations to provide capacitive reactive power to offset the inductive reactive power typical of most power system loads and transmission lines. With long inter-tie transmission lines, series capacitors are used to reduce the effective impedance of the line. By adding thyristors to both of these types of capacitors, actively controlled reactive power is avail-

able using SVCs and TCSC devices, which are shunt-
and series-controlled capacitors, respectively. The thyristors are used to adjust the total impedance of the
device by switching individual modules. Unified
power-flow controllers (UPFCs) also fall into this category.

Objective: FACTS devices are designed to control the flow of
power through the transmission grid.

Benefits: These devices can increase the transfer capacity of
the transmission system, support bus voltages by **F-38**
National Transmission Grid Study providing reactive
power, or be used to enhance dynamic or transient stability.

Barriers: As with HVDC, the power electronics are expensive
and specially trained technicians are needed to maintain
them. In addition, experience is needed to fully understand the coordinated control strategy of these devices
as they penetrate the system.

Commercial Status: As mentioned above, the viability of
HVDC systems has already been demonstrated. American Electric Power (AEP) has installed a FACTS device
in its system, and a new device was recently commissioned by the New York Power Authority (NYPA) to
regulate flows in the northeast.

FACTS Phase-Shifting Transformers: Phase shifters are transformers configured to change the phase angle between
buses; they are particularly useful for controlling the
power flow on the transmission network. Adding thyristor control to the various tap settings of the phase-
shifting transformer permits continuous control of the
effective phase angle (and thus control of power flow).

Objective: Adjust power flow in the system.

Benefits: The key advantage of adding power electronics to
what is currently a non-electronic technology is faster

response time (less then one second vs. about one minute). However, traditional phase shifters still permit redirection of flows and thereby increase transmission system capacity.

Barriers: Traditional phase shifters are deployed today. The addition of the power electronics to these devices is relatively straightforward but increases expense and involves barriers similar to those noted for FACTS compensators.

Commercial Status: Tap-changing phase shifters are available today. Use of thyristor controls is emerging.

FACTS Dynamic Brakes: A dynamic brake is used to rapidly extract energy from a system by inserting a shunt resistance into the network. Adding thyristor controls to the brake permits addition of control functions, such as on-line damping of unstable oscillations.

Objective: Dynamic brakes enhance power system stability. *Benefits*: This device can damp unstable oscillations triggered by equipment outages or system configuration changes.

Barriers: In addition the power electronics issues mentioned earlier, siting a dynamic brake and tuning the device in response to specific contingencies requires careful study.

Commercial Status: BPA has installed a dynamic brake on their system.

Energy-Storage Devices

The traditional function of an energy-storage device is to save production costs by holding cheaply generated off-peak energy that can be dispatched during peak-consumption periods. By virtue of its attributes, energy storage can also provide effective power system control with modest incremental investment. Different dispatch modes can be superimposed on the daily cycle of energy storage, with additional capacity reserved for the express purpose of providing these control functions.

Batteries: Batteries use converters to transform the DC in the storage device to the AC of the power grid. Converters also operate in the opposite direction to recharge the batteries.

Objective: Store energy generated in off-peak hours to be used for emergencies or on-peak needs.

Benefits: Battery converters use thyristors that, by the virtue of their ability to rapidly change the power exchange, can be utilized for a variety of real-time control applications ranging from enhancing transient to preconditioning the area control error for automatic generator control enhancement. During their operational lifetime, batteries have a small impact on the environment. For distributed resources, batteries do not need to be as large as for large-scale generation, and they become important components for regulating micro-grid power and allowing interconnection with the rest of the system.

Barriers: The expense of manufacturing and maintaining batteries has limited their impact in the industry.

Commercial Status: Several materials are used to manufacture batteries though large arrays of lead-acid batteries continue to be the most popular for utility installations. Interest is also growing in so-called "flow batteries" that charge and discharge a working fluid exchanged between two tanks. The emergence of the distributed energy business has increased the interest in deploying batteries for regional energy storage. One of the early battery installations that demonstrated grid benefit was a joint project between EPRI and Southern California Edison at the Chino substation in southern California.

Super-conducting Magnetic Energy Storage (SMES): SMES uses cryogenic technology to store energy by circulating current in a super-conducting coil.

Objective: Store energy generated in off-peak hours to be used for emergencies or on-peak needs.

Benefits: The benefits are similar to those for batteries. SMES devices are efficient because of their super-conductive properties. They are also very compact for the amount of energy stored.

Barriers: As with the super-conducting equipment mentioned in the passive equipment section above, SMES entails costs for the cooling system, the special protection needed in the event the super-conducting device quenches, and the specialized skills required to maintain the device.

Commercial Status: Several SMES units have been commissioned in North America. They have been deployed at Owens Corning to protect plant processes, and at Wisconsin Public Service to address low-voltage and grid instability issues.

Pumped Hydro and Compressed-Air Storage: Pumped hydro consists of large ponds with turbines that can be run in either pump or generation modes. During periods of light load (e.g., night) excess, inexpensive capacity drives the pumps to fill the upper pond. During heavy load periods, the water generates electricity into the grid. Compressed air storage uses the same principle except that large, natural underground vaults are used to store air under pressure during light-load periods.

Objective: This technology helps shave peak and can help in light-load, high-voltage situations.

Benefits: These storage systems behave like conventional generation and have the benefit of producing additional generation sources that can be dispatched to meet various energy and power needs of the system. Air emission issues can be mitigated when base generation is used in off-peak periods as an alternative to potentially high-polluting peaking units during high use periods.

Barriers: Pumped hydro, like any hydro generation project, requires significant space and has corresponding eco-logical impact. The loss of efficiency between pumping and generation as well as the installation and mainte-nance costs must be outweighed by the benefits.

Commercial Status: Pumped hydro projects are sprinkled across North America. A compressed-air storage plant was built in Alabama, and a proposed facility in Ohio may become the world's largest.

Flywheels: Flywheels spin at high velocity to store energy. As with pumped hydro or compressed-air storage, the flywheel is connected to a motor that either accelerates the flywheel to store energy or draws energy to generate electricity. The flywheel rotors are specially designed to significantly reduce losses. Super conductivity technol-ogy has also been deployed to increase efficiency.

Objective: Shave peak energy demand and help in light-load, high-voltage situations. As a distributed resource, fly-wheels enhance power quality and reliability.

Benefits: Flywheel technology has reached low-loss, high-ef-ficiency levels using rotors made of composite materials running in vacuum spaces. Emissions are not an issue for flywheels, except those related to the energy ex-pended to accelerate and maintain the flywheel system.

Barriers: The use of super-conductivity technology faces the same Barriers as noted above under super-conducting cables and SMES. High-energy-storage flywheels re-quire significant space and the high-speed spinning mass can be dangerous if the equipment fails.

Commercial Status: Flywheel systems coupled with batteries are making inroads for small systems (e.g., computer UPS, local loads, electric vehicles). Flywheels rated in the 100 to 200 kW range are proposed for development in the near term.

Controllable Load

Fast-acting load control is an important element in active measures for enhancing the transmission grid. Automatic load shedding (under-frequency, under-voltage), operator-initiated interruptible load, demand-side management programs, voltage reduction, and other load-curtailment strategies have long been an integral part of coping with unforeseen contingencies as a last resort, and/or as a means of assisting the system during high stress, overloaded conditions. Future advances in load-control technology will leverage the advent of real-time pricing, enabling consumers to "back off " their loads (either automatically through grid-friendly appliances or through manual intervention) when the price is right.

Price-Responsive Load: The electricity industry has been characterized by relatively long-term contracts for electricity use. As the industry restructures to be more market-driven, adjusting demand based on market signals will become an important tool for grid operators.

Objective: Inform energy users of system conditions though price signals that nudge consumption into positions that make the system more reliable and economic.

Benefits: The approach reduces the need for new transmission and siting of new generation. Providing incentives to change load in appropriate regions of the system can stabilize energy markets and enhance system reliability. Shifting load from peak periods to less polluting off-peak periods can reduce emissions.

Barriers: The vast number of loads in the system make communication and coordination difficult. Also, using economic signals in real time or near-real time to affect demand usage has not been part of the control structure that has been used by the industry for decades. A common vision and interface standards are needed to coordinate the information exchange required.

Commercial Status: Demand-management programs have been implemented in various areas of the country. These have relied on centralized control. With the advent of the Internet and new distributed information technology approaches, firms are emerging to take advantage of this technology with a more distributed control strategy.

Intelligent Building Systems: Energy can be saved through increasing the efficient operation of buildings and factories. Coordinated utilization of cooling, heating, and electricity in these establishments can significantly reduce energy consumption. Operated in a system that supports price-responsive load, intelligent building systems can benefit system operations. Note: these systems may have their own, local generation. Such systems have the option of selling power to the grid as well as buying power.

Objective: Reduce energy costs and provide energy management resources to stabilize energy markets and enhance system reliability.

Benefits: Such systems optimize energy consumption for the building operators and may provide system operators with energy by reducing load or increasing local generation based on market conditions.

Barriers: These systems require a greater number of sensors and more complex control schemes than are common today. Should energy market access become available at the building level, the price incentives would increase. *Commercial Status*: Pilot projects have been implemented throughout the country.

Generation

Devices that are designed to improve the efficiency or interface of generation resources can be used for power system control. Advanced converter concepts will play an increasing role, providing power conversion between DC and AC power, for resources such as wind, solar, and any nonsynchronous generation. Converter concepts such as pulse width modulation and step-wave inverters would be particularly useful for incorporating DC sources into the grid or providing an asynchronous generation interface. Asynchronous generation has been proposed for increasing the efficiency of hydroelectric generation, which would also have the advantage of providing control functions such as the ability to modify the effective inertia of generators.

Distributed Generation (DG): Fuel cells, micro-turbines, diesel generators, and other technologies are being integrated using power electronics. As these distributed resources increase in number, they can become a significant resource for reliable system operations. Their vast numbers and teaming with local load put them in a similar category to the controllable load discussed above.

Objective: Address local demand cost-effectively.

Benefits: DG is generally easier to site, entails smaller individual financial outlay, and can be more rapidly installation than large-scale generation. DG can supply local load or sell into the system and offers owners self-determination. Recovery and use of waste heat from some DG greatly increases energy efficiency.

Barriers: Volatility of fuel costs and dependence on the fuel delivery infrastructure creates financial and reliability risks. DG units require maintenance and operations expertise, and utilities can set up discouraging rules for interconnection. System operators have so far had difficulty coordinating the impact of DG.

Commercial Status: Deployment of DG units continues to increase. As with controllable load, system operations are recognizing the potential positive implications of DG to stabilize market prices and enhance system reliability though this requires a different way of thinking from the traditional, hierarchical control paradigm.

Real-time Monitoring

This section discusses the impact of new hardware technology on the capacity to sense in real time the loading and limits of individual system devices as well as the overall state of the system. The capability of the electricity grid is restricted through a combination of the limits on individual devices and the composite loadability of the system. Improving monitoring to determine these limits in real time and to measure the system state directly can increase grid capability.

Power-System Device Sensors

The operation of most of the individual devices in a power system (such as transmission lines, cables, transformers, and circuit breakers) is limited by each device's thermal characteristics. In short, trying to put too much power through a device will cause it to heat excessively and eventually fail. Because the limits are thermal, their actual values are highly dependent upon each device's heat dissipation, which is related to ambient conditions. The actual flow of power through most power-system devices is already adequately measured. The need is for improved sensors to dynamically determine the limits by directly or indirectly measuring temperature.

Direct Measurement of Conductor Sag: For overhead transmission lines the ultimate limiting factor is usually conductor sag. As wires heat, they expand, causing the line to sag. Too much sag will eventually result in a short circuit because of arcing from the line to whatever is underneath.

Objective: Dynamically determine line capacity by directly measuring the sag on critical line segments.

Benefits: Dynamically determined line ratings allow for increased power capacity under most operating conditions.

Barriers: Requires continuous monitoring of critical spans. Cost depends on the number of critical spans that must be monitored, the cost of the associated sensor technology, and ongoing cost of communication.

Commercial Status: Pre-commercial units are currently being tested. Approaches include either video or the use of differential GPS. EPRI currently is testing a video-based "sagometer." An alternative is to use differential GPS to directly measure sag. Differential GPS has been demonstrated to be accurate significantly below half a meter.

Indirect Measurement of Conductor Sag: Transmission line sag can also be estimated by physically measuring the conductor temperature using an instrument directly mounted on the line and/or a second instrument that measures conductor tension at the insulator supports.

Objective: Dynamically determine the line capacity.

Benefits: Dynamically determined line ratings allow for increased power capacity under most operating conditions.

Barriers: Requires continuous monitoring of critical spans. Cost depends upon the number of critical spans that must be monitored, the cost of the associated sensor technology, and ongoing costs of communication.

Commercial Status: Commercial units are available.

Indirect Measurement of Transformer Coil Temperature: Similar to transmission line operation, transformer operation is limited by thermal constraints. However, transformers constraints are localized hot spots on the windings that result in breakdown of insulation.

Objective: Dynamically determine transformer capacity. *Benefits*: Dynamically determined transformer ratings allow for increased power capacity under most operating conditions.

Barriers: The simple use of oil temperature measurements is usually considered to be unreliable.

Commercial Status: Sophisticated monitoring tools are now commercially available that combine several different temperature and current measurements to dynamically determine temperature hot spots.

Underground/Submarine Cable Monitoring/Diagnostics: The below-surface cable systems described above require real-time monitoring to maximize their use and warn of potential failure.

Objective: Incorporate real-time sensing equipment to detect potentially hazardous operating situations as well as dynamic limits for safe flow of energy.

Benefits: Monitoring equipment maximizes the use of the transmission asset, mitigates the risk of failure and the ensuing expense of repair, and supports preventive maintenance procedures. The basic sensing and monitoring technology is available today.

Barriers: The level of sophistication of the sensing and monitoring equipment adds to the cost of the cable system.

The use of dynamic limits must also be integrated into system operation procedures and the associated tools of existing control facilities.

Commercial Status: Newer cable systems are being designed with monitoring/diagnostics in mind. Cable temperature, dynamic thermal rating calculations, partial discharge detection, moisture ingress, cable damage, hydraulic condition (as appropriate), and loss detection are some of the sensing functions being put in place. Multifunctional cables are also being designed and deployed (particularly submarine cables) that include communications capabilities. Monitoring is being integrated directly into the manufacturing process of these cables.

Direct System-State Sensors

In some situations, transmission capability is not limited by individual devices but rather by region-wide dynamic loadability constraints. These include transient stability limitations, oscillatory stability limitations, and voltage stability limitations. Because the time frame associated with these phenomena is much shorter than that associated with thermal overloads, predicting, detecting and responding to these events requires much faster real-time state sensors than for thermal conditions. The system state is characterized ultimately by the voltage magnitudes and angles at all the system buses. The goal of these sensors is to provide these data at a high sampling rate.

Power-System Monitors

Objective: Collect essential signals (key power flows, bus voltages, alarms, etc.) from local monitors available to site operators, selectively forwarding to the control center or to system analysts.

Benefits: Provides regional surveillance over important parts of the control system to verify system performance in real time.

Barriers: Existing SCADA and Energy Management Systems provide low-speed data access for the utility's infrastructure. Building a network of high-speed data monitors with intra-regional breadth requires collaboration among utilities within the interconnected power system.

Commercial Status: BPA has developed a network of dynamic monitors collecting high-speed data, first with the power system analysis monitor (PSAM), and later with the portable power system monitor (PPSM), both early examples of WAMS products.

Phasor Measurement Units (PMUs)

Objective: PMUs are synchronized digital transducers that can stream data, in real time, to phasor data concentrator (PDC) units. The general functions and topology for this network resemble those for dynamic monitor networks. Data quality for phasor technology appears to be very high, and secondary processing of the acquired phasors can provide a broad range of signal types.

Benefits: Phasor networks have best value in applications that are mission critical and that involve truly wide-area measurements.

Barriers: Establishing PMU networks is straightforward and has already been done. The primary impediment is cost and assuring value for the investment (making best use of the data collected).

Commercial Status: PMU networks have been deployed at several utilities across the country.

Index

For Product Safety Concerns and Information please contact our EU
representative GPSR@taylorandfrancis.com Taylor & Francis Verlag GmbH,
Kaufingerstraße 24, 80331 München, Germany

Printed and bound by CPI Group (UK) Ltd, Croydon, CR0 4YY

01/05/2025
01858500-0001